THE MAGIC OF REALITY

www.transworldbooks.co.uk

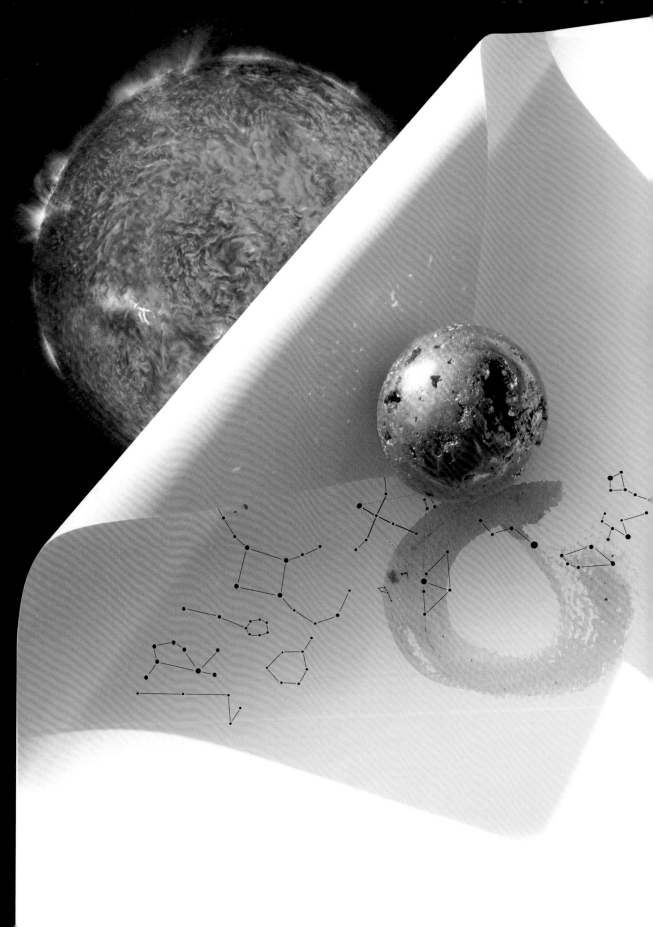

RICHARD DAWKINS

The Magic of Reality

How we know what's really true

ILLUSTRATED BY

DAVE McKEAN

BANTAM PRESS

LONDON · TORONTO · SYDNEY · AUCKLAND · JOHANNESBURG

61–63 Uxbridge Road, London W5 5SA
A Random House Group Company
www.rbooks.co.uk

First published in Great Britain
in 2011 by Bantam Press
an imprint of Transworld Publishers

A CIP catalogue record for this book
is available from the British Library.

ISBN 9780593066126

Addresses for Random House Group Ltd companies outside the UK
can be found at: www.randomhouse.co.uk
The Random House Group Ltd Reg. No. 954009

The Random House Group Limited supports the Forest Stewardship Council (FSC®), the
leading international forest-certification organization. Our books carrying the FSC label are
printed on FSC®-certified paper. FSC is the only forest-certification scheme endorsed by
the leading environmental organizations, including Greenpeace. Our paper-procurement
policy can be found at www.randomhouse.co.uk/environment.

Typeset in Minion
Printed and bound in Germany

4 6 8 10 9 7 5

Clinton John Dawkins

1915–2010

O, my beloved father

Contents

1 What is REALITY? What is MAGIC?

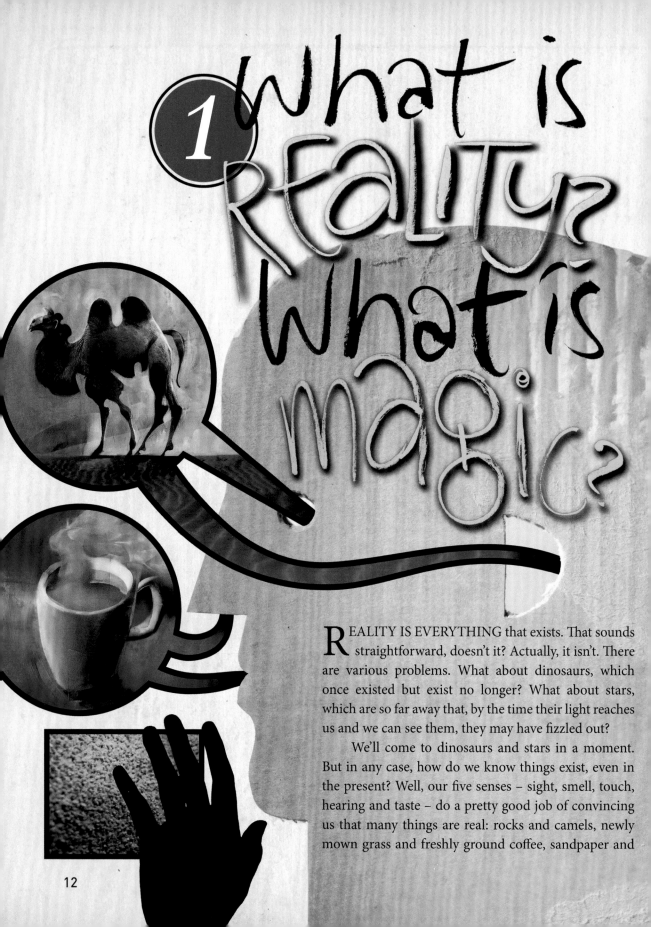

REALITY IS EVERYTHING that exists. That sounds straightforward, doesn't it? Actually, it isn't. There are various problems. What about dinosaurs, which once existed but exist no longer? What about stars, which are so far away that, by the time their light reaches us and we can see them, they may have fizzled out?

We'll come to dinosaurs and stars in a moment. But in any case, how do we know things exist, even in the present? Well, our five senses – sight, smell, touch, hearing and taste – do a pretty good job of convincing us that many things are real: rocks and camels, newly mown grass and freshly ground coffee, sandpaper and velvet, waterfalls and doorbells, sugar and salt. But are we only going to call something 'real' if we can detect it directly with one of our five senses?

What about a distant galaxy, too far away to be seen with the naked eye? What about a bacterium, too small to be seen without a powerful microscope? Must we say that these do not exist because we can't see them? No. Obviously we can enhance our senses through the use of special instruments: telescopes for the galaxy, microscopes for bacteria. Because we understand telescopes and microscopes, and how they work, we can use them to extend the reach of our senses – in this case, the sense of sight – and what they enable us to see convinces us that galaxies and bacteria exist.

How about radio waves? Do they exist? Our eyes can't detect them, nor can our ears, but again special instruments – television sets, for example – convert them into signals that we can see and hear. So, although we can't see or hear radio waves, we know they are a part of reality. As with telescopes and microscopes, we understand how radios and televisions work. So they help our senses to build a picture of what exists: the real world – reality. Radio telescopes (and X-ray telescopes) show us stars and galaxies through what seem like different eyes: another way to expand our view of reality.

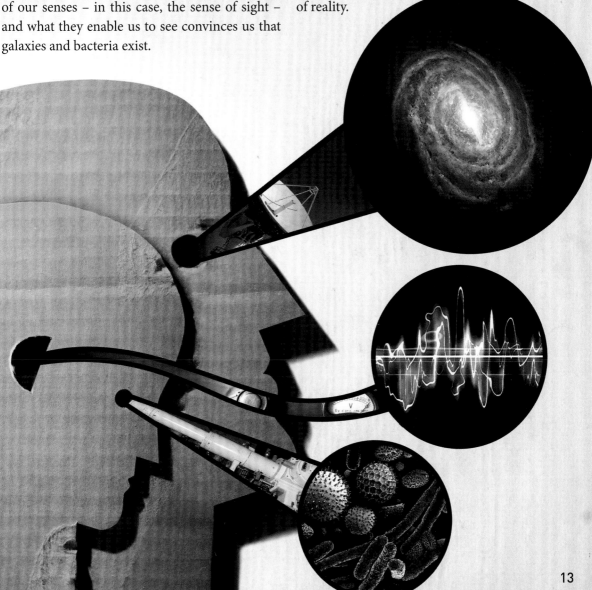

Back to those dinosaurs. How do we know that they once roamed the Earth? We have never seen them or heard them or had to run away from them. Alas, we don't have a time machine to show them to us directly. But here we have a different kind of aid to our senses: we have fossils, and we can see *them* with the naked eye. Fossils don't run and jump but, because we understand how fossils are formed, they can tell us something of what happened millions of years ago. We understand how water, with minerals dissolved in it, seeps into corpses buried in layers of mud and rock. We understand how the minerals crystallize out of the water and replace the materials of the corpse, atom by atom, leaving some trace of the original animal's form imprinted on the stone. So, although we can't see dinosaurs directly with our senses, we can work out that they must have existed, using indirect evidence that still ultimately reaches us through our senses: we see and touch the stony traces of ancient life.

In a different sense, a telescope can work like a kind of time machine. What we see when we look at anything is actually light, and light takes time to travel. Even when you look at a friend's face you are seeing them in the past, because the light from their face takes a tiny fraction of a second to travel to your eye. Sound travels much more slowly, which is why you see a firework burst in the sky noticeably earlier than you hear the bang. When you watch a man chopping down a tree in the distance, there is an odd delay in the sound of his axe hitting the tree.

Light travels so fast that we normally assume anything we see happens at the instant we see it. But stars are another matter. Even the sun is eight light-minutes away. If the sun blew up, this catastrophic event wouldn't become a part of our reality until eight minutes later. And that would be the end of us! As for the next nearest star, Proxima Centauri, if you look at it in 2011, what you are seeing is happening in 2007. Galaxies are huge collections of stars. We are in one galaxy called the Milky Way. When you look at the Milky Way's next-door neighbour, the Andromeda galaxy, your telescope is a time machine taking you back two and a half million years. There's a cluster of five galaxies called Stephan's Quin-tet, which we see through the Hubble telescope – spectacularly colliding with each other. But we see them colliding 280 million years ago. If there are aliens in one of those colliding galaxies with a tele-scope powerful enough to see us, what they are seeing

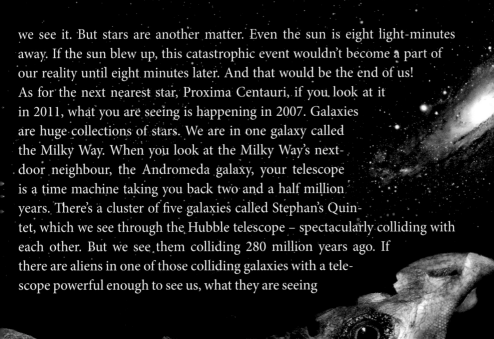

on Earth, at this very moment, here and now, is the early ancestors of the dinosaurs. Are there really aliens in outer space? We've never seen or heard them. Are they a part of reality? Nobody knows; but we do know what kind of things could one day tell us if they are. If ever we got near to an alien, our sense organs could tell us about it. Perhaps somebody will one day invent a telescope powerful enough to detect life on other planets from here. Or perhaps our radio telescopes will pick up messages that could only have come from an alien intelligence. For reality doesn't just consist of the things we already know about: it also includes things that exist but that we don't know about yet – and won't know about until some future time, perhaps when we have built better instruments to assist our five senses.

Atoms have always existed, but it was only rather recently that we became sure of their existence, and it is likely that our descendants will know about many more things that, for now, we do not. That is the wonder and the joy of science: it goes on and on uncovering new things. This doesn't mean we should believe just *anything* that anybody might dream up: there are a million things we can imagine but which are highly unlikely to be real – fairies and hobgoblins, leprechauns and hippogriffs. We should always be open-minded, but the only good reason to believe that something exists is if there is real evidence that it does.

Models: testing our imagination

There is a less familiar way in which a scientist can work out what is real when our five senses cannot detect it directly. This is through the use of a 'model' of what *might* be going on, which can then be tested. We imagine – you might say we guess – what might be there. That is called the model. We then work out (often by doing a mathematical calculation) what we ought to see, or hear, etc. (often with the help of measuring instruments) if the model were true. We then check whether that is what we actually do see. The model might literally be a replica made out of wood or plastic, or it might be a piece of mathematics on paper, or it might be a *simulation* in a computer. We look carefully at the model and *predict* what we ought to see (hear, etc.) with our senses (with the aid of instruments, perhaps) if the model were correct. Then we look to see whether the predictions are right or wrong. If they are right, this increases our confidence that the model really does represent reality; we then go on to devise further experiments, perhaps refining the model, to test the findings further and confirm them. If our predictions are wrong, we reject the model, or modify it and try again.

Here's an example. Nowadays, we know that genes – the units of heredity – are made of stuff called DNA. We know a great deal about DNA and how it works. But you can't see the details of what DNA looks like, even with a powerful microscope. Almost everything we know about DNA comes indirectly from dreaming up models and then testing them.

Actually, long before anyone had even heard of DNA, scientists already knew lots about genes from testing the predictions of models. Back in the nineteenth century, an Austrian monk called Gregor Mendel did experiments in his monastery garden, breeding peas in large quantities. He counted the numbers of plants that had flowers of various colours, or that had peas that were wrinkly or smooth, as the generations went by. Mendel never saw or touched a gene. All he saw were peas and flowers, and he could use his eyes to *count* different types. He invented a *model*, which involved what we would now call genes (though Mendel didn't

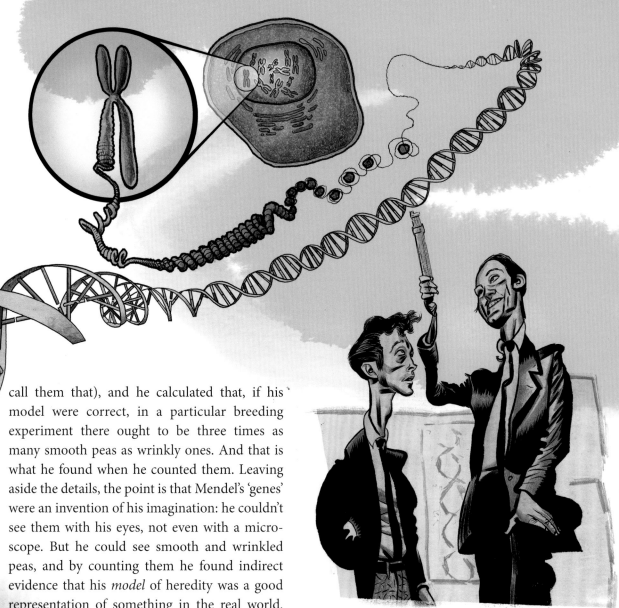

call them that), and he calculated that, if his model were correct, in a particular breeding experiment there ought to be three times as many smooth peas as wrinkly ones. And that is what he found when he counted them. Leaving aside the details, the point is that Mendel's 'genes' were an invention of his imagination: he couldn't see them with his eyes, not even with a microscope. But he could see smooth and wrinkled peas, and by counting them he found indirect evidence that his *model* of heredity was a good representation of something in the real world. Later scientists used a modification of Mendel's method, working with other living things such as fruit flies instead of peas, to show that genes are strung out in a definite order, along threads called chromosomes (we humans have forty-six chromosomes, fruit flies have eight). It was even possible to work out, by testing models, the exact order in which genes were arranged along chromosomes. All this was done long before we knew that genes were made of DNA.

Nowadays we know this, and we know exactly how DNA works, thanks to James Watson and Francis Crick, plus a lot of other scientists who came after them. Watson and Crick could not see DNA with their own eyes. Once again, they made their discoveries by imagining models and testing them. In their case, they literally built metal and cardboard models of what DNA might look like, and they calculated what certain measurements ought to be if those models were correct. The predictions of one model, the so-called double helix model, exactly fitted the measurements made by Rosalind Franklin and Maurice Wilkins, using special instruments

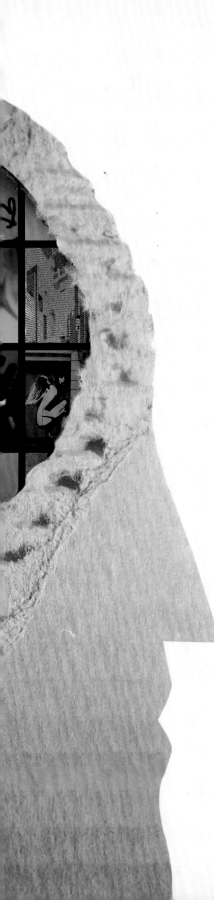

involving X-rays beamed into crystals of purified DNA. Watson and Crick also immediately realized that their model of the structure of DNA would produce exactly the kind of results seen by Gregor Mendel in his monastery garden.

We come to know what is real, then, in one of three ways. We can detect it directly, using our five senses; or indirectly, using our senses aided by special instruments such as telescopes and microscopes; or even more indirectly, by creating models of what *might* be real and then testing those models to see whether they successfully predict things that we can see (or hear, etc.), with or without the aid of instruments. Ultimately, it always comes back to our senses, one way or another.

Does this mean that reality only contains things that can be detected, directly or indirectly, by our senses and by the methods of science? What about things like jealousy and joy, happiness and love? Are these not also real?

Yes, they are real. But they depend for their existence on brains: human brains, certainly, and probably the brains of other advanced animal species, such as chimpanzees, dogs and whales, too. Rocks don't feel joy or jealousy, and mountains do not love. These emotions are intensely real to those who experience them, but they didn't exist before brains did. It is possible that emotions like these – and perhaps other emotions that we can't begin to dream of – could exist on other planets, but only if those planets also contain brains – or something equivalent to brains: for who knows what weird thinking organs or feeling machines may lurk elsewhere in the universe?

Science and the supernatural: explanation and its enemy

So that is reality, and that is how we can know whether something is real or not. Each chapter of this book is going to be about one particular aspect of reality – the sun, for instance, or earthquakes, or rainbows, or the many different kinds of animals. I want now to turn to the other key word of my title: magic. Magic is a slippery word: it is commonly used in three different ways, and the first thing I must do is distinguish between them. I'll call the first one 'supernatural magic', the second one 'stage magic' and the third one (which is my favourite meaning, and the one I intend in my title) 'poetic magic'.

Supernatural magic is the kind of magic we find in myths and fairy tales. (In 'miracles', too, though I shall leave those to one side for now and return to them in the final chapter.) It's the magic of Aladdin's lamp, of wizards' spells, of the Brothers Grimm, of Hans Christian Andersen and of J. K. Rowling. It's the fictional magic of a witch casting a spell and turning a prince into a frog, or a fairy god-mother changing a pumpkin into a gleaming coach. These are the stories we all remember with fondness from our childhood, and many of us still enjoy when served up in a traditional Christmas pantomime – but we all know this kind of magic is just fiction and does not happen in reality.

Stage magic, by contrast, really does happen, and it can be great fun. Or at least, *something* really happens, though it isn't what the audience thinks it is. A man on a stage (it usually is a man, for some reason, so I shall say 'he' but you can insert 'she' if you prefer) deceives us into thinking that something astonishing (it may even *seem* supernatural) has happened when what *really* happened was something quite different. Silk hand-kerchiefs cannot turn into rabbits, any more than frogs can turn into princes. What we have seen on the stage is only a trick. Our eyes have deceived us – or rather, the conjuror has gone to great pains to deceive our eyes, perhaps by cleverly using words to distract us from what he is really doing with his hands.

Some conjurors are honest and go out of their way to make sure their audiences know that they have simply performed a trick. I am

> *What Number am I thinking of?*

thinking of people like James 'The Amazing' Randi, or Penn and Teller, or Derren Brown. Even though these admirable performers don't usually tell the audience exactly *how* they did the trick – they could be thrown out of the Magic Circle (the conjurors' club) if they did that – they do make sure the audience knows that there was no supernatural magic involved. Others don't actively spell out that it was just a trick, but they don't make exaggerated claims about what they have done either – they just leave the audience with the rather enjoyable sensation that something mysterious has happened, without actively lying about it. But unfortunately there are some conjurors who are deliberately dishonest, and who pretend they really do have 'super-natural' or 'paranormal' powers: perhaps they claim that they really can bend metal or stop clocks by the power of thought alone. Some of these dishonest fakes ('charlatans' is a good word for them) earn large fees from mining or oil companies by claiming that they can tell, using 'psychic powers', where would be a good place to drill. Other charlatans exploit people who are grieving, by claiming to be able to make contact with the dead. When this happens it is no longer just fun or entertainment, but preying on people's gullibility and distress. To be fair, it may be that not all of these people are charlatans. Some of them may sincerely believe they are talking to the dead.

The third meaning of magic is the one I mean in my title: poetic magic. We are moved to tears by a beautiful piece of music and we describe the performance as 'magical'. We gaze up at the stars on a dark night with no moon and no city lights and, breathless with joy, we say the sight is 'pure magic'. We might use the same word to describe a gorgeous sunset, or an alpine landscape, or a rainbow against a dark sky. In this sense, 'magical' simply means deeply moving, exhilarating: something that gives us goose bumps, something that makes us feel more fully alive. What I hope to show you in this book is that reality – the facts of the real world as understood through the methods of science – is magical in this third sense, the poetic sense, the good-to-be-alive sense.

Now I want to return to the idea of the supernatural and explain why it can never offer us a true explanation of the things we see in the world and universe around us. Indeed, to claim a supernatural explanation of something is not to explain it at all and, even worse, to rule out any possibility of its ever being explained. Why do I say that? Because anything 'supernatural' must by definition be beyond the reach of a natural explanation. It must be beyond the reach of science and the well-established, tried and tested scientific method that has been responsible for the huge advances in knowledge we have enjoyed over the last 400 years or so. To say that something happened supernaturally is not just to say 'We don't understand it' but to say 'We will never understand it, so don't even try'.

Science takes exactly the opposite approach. Science thrives on its inability – so far – to explain everything, and uses that as the spur to go on asking questions, creating possible models and testing them, so that we make our way, inch by inch, closer to the truth. If something were to happen that went against our current understanding of reality, scientists would see that as a challenge to our present model, requiring us to abandon or at least change it. It is through such adjustments and subsequent testing that we approach closer and closer to what is true.

What would you think of a detective who, baffled by a murder, was too lazy even to try to work at the problem and instead wrote the mystery off as 'supernatural'? The whole history of science shows us that things once thought to be the result of the supernatural – caused by gods (both happy and angry), demons, witches, spirits, curses and spells – actually do have natural explanations: explanations that we can understand and test and have confidence in. There is absolutely no reason to believe that those things for which science does not *yet* have natural explanations will turn out to be of supernatural origin, any more than volcanoes or earthquakes or diseases turn out to be caused by angry deities, as people once believed they were.

Of course, no one really believes that it would be possible to turn a frog into a prince (or was it a prince into a frog? I can never remember) or a pumpkin into a coach, but have you ever stopped to consider *why* such things would be impossible? There are various ways of explaining it. My favourite way is this.

Frogs and coaches are complicated things, with lots of parts that need to be put together in a special way, in a special pattern that can't just happen by accident (or by a wave of a wand). That's what 'complicated' means. It is very difficult to make a complicated thing like a frog or a coach. To make a coach you need to bring all the parts together in just the right way. You need the skills of a carpenter and other craftsmen. Coaches don't just happen by chance or by snapping your fingers and saying 'Abracadabra'. A coach has structure, complexity, working parts: wheels and axles, windows and doors, springs

example 1   *fig.1*   *fig.2*   *fig.3*   *fig.4*

and padded seats. It would be relatively easy to turn something complicated like a coach into something simple – like ash, for instance: the fairy godmother's wand would just need a built-in blowtorch. It is easy to turn almost anything into ash. But no one could take a pile of ash – or a pumpkin – and turn it into a coach, because a coach is too complicated; and not just complicated, but complicated *in a useful direction*: in this case useful for people to travel in.

Let's make it a bit easier for the fairy god-mother by supposing that, instead of calling for a pumpkin, she had called for all the *parts* you need for assembling a coach, all jumbled together in a box, like a kit for building a model plane. The kit for making a coach consists of hundreds of planks of wood, panes of glass, rods and bars of iron, wads of padding and sheets of leather, along with nails, screws and pots of glue to hold things together. Now suppose that, instead of reading the instructions and joining the parts in an orderly sequence, she just put all the bits into a great big bag and shook them up. What are the chances that the parts would happen to

stick themselves together in just the right way to assemble a working coach? The answer is – effectively zero. And a part of the reason for that is the massive number of *possible* ways in which you could combine the shuffled bits and pieces which would not result in a working coach – or a working *anything*.

If you take a load of parts and shake them around at random, they may just occasionally fall into a pattern that is useful, or that we otherwise recognize as somehow special. But the number of ways in which that can happen is tiny: very tiny indeed compared with the number of ways in which they will fall into a pattern that we don't recognize as anything more than a heap of junk. There are millions of ways of shuffling and reshuffling a heap of bits and pieces: millions of ways of transforming them into . . . another heap of bits and pieces. Every time you shuffle them, you get a unique heap of junk that has never been seen before – but only a tiny minority of those millions of possible heaps will do anything useful (such as taking you to the ball) or will be remarkable or memorable in any way.

example 2

fig.1

fig.2

fig.3

Sometimes we can literally count the number of ways you can reshuffle a series of bits – as with a pack of cards, for instance, where the 'bits' are the individual cards.

Suppose the dealer shuffles the pack and deals them out to four players, so that they each have 13 cards. I pick up my hand and gasp in astonishment. I have a complete hand of *13 spades*! All the spades.

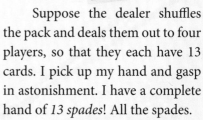

I am too startled to go on with the game and I show my hand to the other three players, knowing they will be as amazed as I am.

But then, one by one, each of the other players lays his cards on the table, and the gasps of astonishment grow with each hand. Every one of them has a 'perfect' hand: one has 13 hearts, another has 13 diamonds, and the last one has 13 clubs.

Would this be supernatural magic? We might be tempted to think so. Mathematicians can calculate the chance of such a remarkable deal happening purely by chance. It turns out to be almost impossibly small: 1 in 53,644,737,765,488,792,839,237,440,000. I'm not sure I would even know how to say that number! If you sat down and played cards for a trillion years, you might on one occasion get a perfect deal like that. But – and here's the thing – this deal is no more unlikely than *every other deal of cards that has ever happened!* The chance of *any* particular deal of 52 cards is 1 in 53,644,737,765,488,792,839,237,440,000 because that is the total number of all possible deals. It is just that we don't notice any particular pattern in the vast majority of deals that are made, so they don't strike us as anything out of the ordinary. We only notice the deals that happen to stand out in some way.

There are billions of things you could turn a prince into, if you were brutal enough to rearrange his bits into billions of combinations at random. But most of those combinations would look like a mess – like all those billions of meaningless, random hands of cards that have been dealt. Only a tiny minority of those possible combinations of randomly shuffled prince-bits would be recognizable or good for anything at all, let alone a frog.

Princes don't turn into frogs, and pumpkins don't turn into coaches, because frogs and coaches are complicated things whose bits could have been combined into a near-infinite number of heaps of junk. And yet we know, as a fact, that every living thing – every human, every crocodile, every blackbird, every tree and even every Brussels sprout – has evolved from other, originally simpler forms. So isn't *that* just a process of luck, or a kind of magic? No! Absolutely not! This is a very common misunderstanding, so I want to explain right now why what we see in real life is

not the result of chance or luck or anything remotely 'magical' at all (except, of course, in the strictly poetic sense of something that fills us with awe and delight).

The slow magic of evolution

To turn one complex organism into another complex organism in a single step – as in a fairytale – would indeed be beyond the realms of realistic possibility. And yet complex organisms *do* exist. So how did they arise? How, in reality, did complicated things like frogs and lions, baboons and banyan trees, princes and pumpkins, you and me come into existence? For most of history that was a baffling question, which no one could answer properly. People therefore invented stories to try to explain it. But then the question was answered – and answered brilliantly – in the nineteenth century, by one of the greatest scientists who ever lived, Charles Darwin. I'll use the rest of this chapter to explain his answer, briefly, and in different words from Darwin's own.

The answer is that complex organisms – like humans, crocodiles and Brussels sprouts – did not come about suddenly, in one fell swoop, but gradually, step by tiny step, so that what was there after each step was only a little bit different from what was already there before. Imagine you wanted to create a frog with long legs. You could give yourself a good start by beginning with something that was already a bit like what you wanted to achieve: a frog with short legs, say. You would look over your short-legged frogs and measure their legs. You'd pick a few males and a few females that had slightly longer legs than most, and you'd let them mate together, while preventing their shorter-legged friends from mating at all.

The longer-legged males and females would make tadpoles together, and these would eventually grow legs and become frogs. Then you'd measure this new generation of frogs, and once again pick out those males and females that had longer-than-average legs, and put them together to mate.

After doing this for about 10 generations, you might start to notice something interesting. The average leg length of your population of frogs would now be noticeably longer than the average leg length of the starting population. You might even find that *all* the frogs of the 10th generation had longer legs than any of the frogs of the first generation. Or 10 generations might not be enough to achieve this: you might need to go on for 20 generations or even more. But eventually you could proudly say, 'I have made a new kind of frog with longer legs than the old type.'

No wand was needed. No magic of any kind was required. What we have here is the process called *selective breeding*. It makes use of the fact that frogs vary among themselves and those variations tend to be inherited – that is, passed on from parent to child via the genes. Simply by

choosing which frogs breed and which do not, we can make a new kind of frog.

Simple, isn't it? But just making legs longer is not very impressive. After all, we started with frogs – they were just short-legged frogs. Suppose you started, not with a shorter-legged form of frog, but with something that wasn't a frog at all, say something more like a newt. Newts have very short legs compared with frogs' legs (compared with frogs' *hind* legs, at least), and they use them not for jumping but for walking. Newts also have long tails, whereas frogs don't have tails at all, and newts are altogether longer and narrower than most frogs. But I think you can see that, given enough thousands of generations, you could change a population of newts into a population of frogs, simply by patiently choosing, in each of those millions of generations, male and female newts that were slightly more frog-like and letting them mate together, while preventing their less frog-like friends from doing so. At no stage during the process would you see any dramatic change. Every generation would look pretty much like the previous generation, but nevertheless, once enough generations had gone by, you'd start to notice that the average tail length was slightly shorter and the average pair of hind legs was slightly longer. After a very large number of generations, the longer-legged, shorter-tailed individuals might find it easier to start using their long legs for hopping instead of crawling. And so on.

Of course, in the scenario I have just described, we are imagining ourselves as breeders, picking out those males and females that we want to mate together in order to achieve an end result that *we* have chosen. Farmers have been applying this technique for thousands of years, to produce cattle and crops that have higher yields or are more resistant to disease, and so

29

on. Darwin was the first person to understand that it works *even when there is no breeder to do the choosing*. Darwin saw that the whole thing would happen *naturally*, as a matter of course, for the simple reason that some individuals survive long enough to breed and others don't; and those that survive do so because they are better equipped than others. So the survivors' children inherit the genes that helped their parents to survive. Whether it's newts or frogs, hedgehogs or dandelions, there will always be some individuals that are better at surviving than others. If long legs happen to be helpful (for frogs or grasshoppers jumping out of danger, say, or for cheetahs hunting gazelles or gazelles fleeing from cheetahs), the individuals with longer legs will be less likely to die. They will be more likely to live long enough to reproduce. Also, more of the individuals available for mating with will have long legs. So in every generation there will be a greater chance of the genes for longer legs being passed into the next generation. Over time we will find that more and more of the individuals within that population have the genes for longer

legs. So the effect will be exactly the same as if an intelligent designer, such as a human breeder, had chosen long-legged individuals for breeding – except that *no such designer is required*: it all happens naturally, all by itself, as the automatic consequence of which individuals survive long enough to reproduce, and which don't. For this reason, the process is called *natural selection*.

Given enough generations, ancestors that look like newts can change into descendants that look like frogs. Given even more generations, ancestors that look like fish can change into descendants that look like monkeys. Given yet more generations, ancestors that look like bacteria can change into descendants that look like humans. And this is exactly what happened. This is the kind of thing that happened in the history of every animal and plant that has ever lived. The number of generations required is larger than you or I can possibly imagine, but the world is thousands of millions of years old, and we know from fossils that life got started more than 3,500 million (3.5 billion) years ago, so there has been plenty of time for evolution to happen.

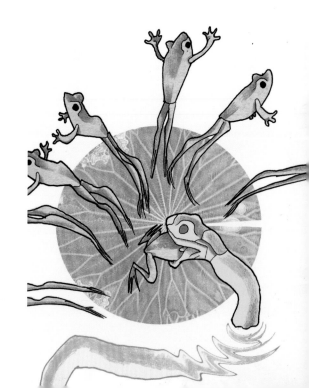

This is Darwin's great idea, and it is called Evolution by Natural Selection. It is one of the most important ideas ever to occur to a human mind. It explains everything we know about life on Earth. Because it is so important, I'll come back to it in later chapters. For now, it is enough to understand that evolution is very slow and gradual. In fact, it is the gradualness of evolution that allows it to make complicated things like frogs and princes. The magical changing of a frog into a prince would be not gradual but sudden, and this is what rules such things out of the world of reality. Evolution is a real explanation, which really works, and has real evidence to demonstrate the truth of it; anything that suggests that complicated life forms appeared suddenly, in one go (rather than evolving gradually step by step), is just a lazy story – no better than the fictional magic of a fairy godmother's wand.

As for pumpkins turning into coaches, magic spells are just as certainly ruled out for them as they are for frogs and princes. Coaches don't evolve – or at least, not naturally, in the same way that frogs and princes do. But coaches – along with airliners and pickaxes, computers and flint arrowheads – are made by humans who *did* evolve. Human brains and human hands evolved by natural selection, just as surely as newts' tails and frogs' legs did. And human brains, once they had evolved, were able to design and create coaches and cars, scissors and symphonies, washing machines and watches. Once again, no magic. Once again, no trickery. Once again, everything beautifully and simply explained.

In the rest of this book I want to show you that the real world, as understood scientifically, has magic of its own – the kind I call poetic magic: an inspiring beauty which is all the more magical because it is real and because we can understand how it works. Next to the true beauty and magic of the real world, supernatural spells and stage tricks seem cheap and tawdry by comparison. The magic of reality is neither supernatural nor a trick, but – quite simply – wonderful. Wonderful, and real. Wonderful *because* real.

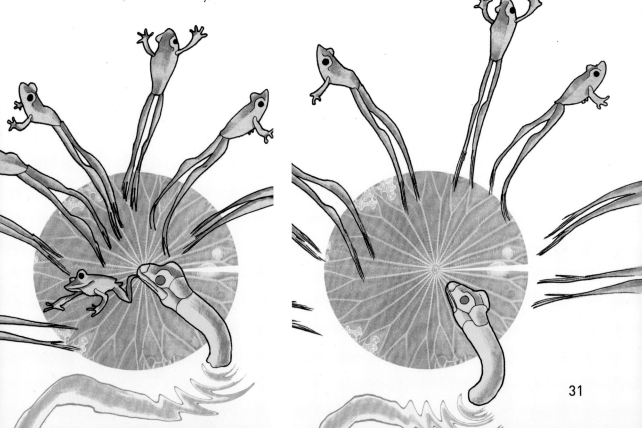

2 WHO WAS the first

MOST CHAPTERS in this book are headed by a question. My purpose is to answer the question, or at least give the best possible answer, which is the answer of science. But I shall usually begin with some mythical answers because they are colourful and interesting, and real people have believed them. Some people still do.

All peoples around the world have origin myths, to account for where they came from. Many tribal origin myths talk only about that one particular tribe – as though other tribes don't count! In the same way, many tribes have a rule that they mustn't kill people – but 'people' turns out to mean only others of your own tribe. Killing members of other tribes is just fine!

Here's a typical origin myth, from a group of Tasmanian aborigines. A god called Moinee was defeated by a rival god called Dromerdeener in a terrible battle up in the stars. Moinee fell out of the stars down to Tasmania to die. Before he died, he wanted to give a last blessing to his final resting place, so he decided to create humans. But he was in such a hurry, knowing he was dying, that he forgot to give them knees; and (no doubt distracted by his plight) he absent-mindedly gave them big tails like kangaroos, which meant they couldn't sit down. Then he died. The people hated having kangaroo tails and no knees, and they cried out to the heavens for help.

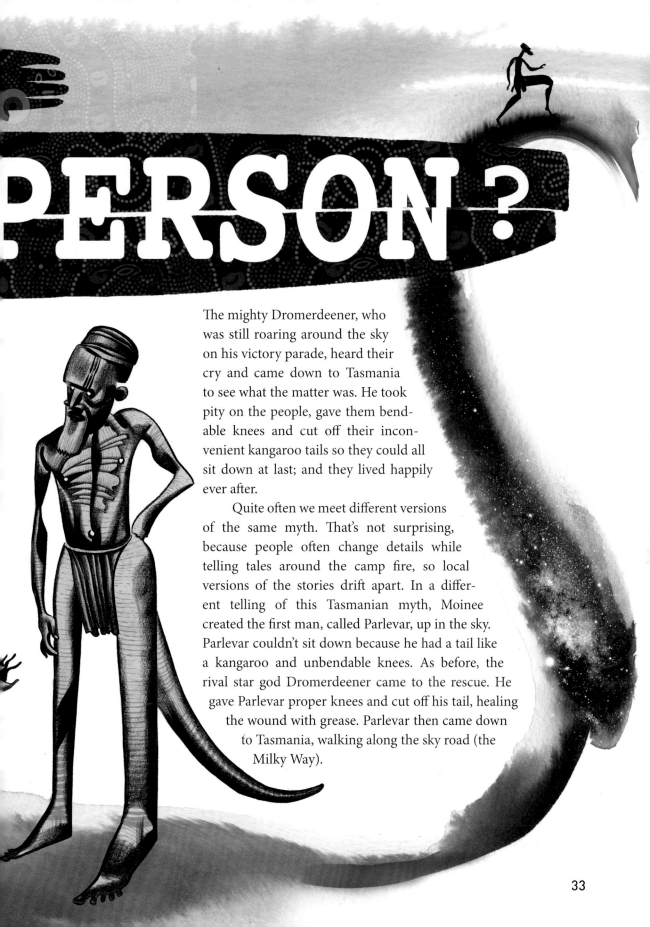

PERSON?

The mighty Dromerdeener, who was still roaring around the sky on his victory parade, heard their cry and came down to Tasmania to see what the matter was. He took pity on the people, gave them bendable knees and cut off their inconvenient kangaroo tails so they could all sit down at last; and they lived happily ever after.

Quite often we meet different versions of the same myth. That's not surprising, because people often change details while telling tales around the camp fire, so local versions of the stories drift apart. In a different telling of this Tasmanian myth, Moinee created the first man, called Parlevar, up in the sky. Parlevar couldn't sit down because he had a tail like a kangaroo and unbendable knees. As before, the rival star god Dromerdeener came to the rescue. He gave Parlevar proper knees and cut off his tail, healing the wound with grease. Parlevar then came down to Tasmania, walking along the sky road (the Milky Way).

The Hebrew tribes of the Middle East had only a single god, whom they regarded as superior to the gods of rival tribes. He had various names, none of which they were allowed to say. He made the first man out of dust and called him Adam (which just means 'man'). He deliberately made Adam like himself. Indeed, most of the gods of history were portrayed as men (or sometimes women), often of giant size and always with supernatural powers.

The god placed Adam in a beautiful garden called Eden, filled with trees whose fruit Adam was encouraged to eat – with one exception. This forbidden tree was the 'tree of knowledge of good and evil', and the god left Adam in no doubt that he must never eat its fruit.

The god then realized that Adam might be lonely all by himself, and wanted to do something about it. At this point – as with the story of Dromerdeener and Moinee – there are two versions of the myth, both found in the biblical book of Genesis. In the more colourful version, the god made all the animals as Adam's helpers, then decided that there was still something missing: a woman! So he gave Adam a general anaesthetic, cut him open, removed one rib and stitched him up again. Then he grew a woman from the rib, rather as you grow a flower from a cutting. He named her Eve and presented her to Adam as his wife.

Unfortunately, there was a wicked snake in the garden, who approached Eve and persuaded her to give Adam the forbidden fruit from the tree of knowledge of good and evil. Adam and Eve ate the fruit and promptly acquired the knowledge that they were naked.

This embarrassed them, and they made themselves aprons out of fig leaves. When the god noticed this he was furious with them for eating the fruit and acquiring knowledge – losing their innocence, I suppose. He threw them out of the garden, and condemned them and all their descendants to a life of hardship and pain. To this day, the story of Adam's and Eve's terrible disobedience is still taken seriously by many people under the name of 'original sin'. Some people even believe we have all inherited this 'original sin' from Adam (although many of them admit that Adam never actually existed!), and share in his guilt.

The Norse peoples of Scandinavia, famous as Viking seafarers, had lots of gods, as the Greeks and Romans did. The name of their chief god was Odin, sometimes called Wotan or Woden, from which we get our 'Wednesday'. ('Thursday' comes from another Norse god, Thor, the god of thunder, which he made with his mighty hammer.)

One day Odin was walking along the seashore with his brothers, who were also gods, and they came upon two tree trunks.

36

One of these tree trunks they turned into the first man, whom they called 'Ask', and the other they turned into the first woman, naming her 'Embla'. Having created the bodies of the first man and first woman, the brother gods then gave them the breath of life, followed by consciousness, faces and the gift of speech.

Why tree trunks, I wonder? Why not icicles or sand dunes? Isn't it fascinating to wonder who made such stories up, and why? Presumably the original inventors of all these myths knew they were fiction at the moment when they made them up. Or do you think many different people came up with different parts of the stories, at different times and in different places, and other people later put them together, perhaps changing some of them, without realizing that the various bits were originally just made up?

Stories are fun, and we all love repeating them. But when we hear a colourful story, whether it is an ancient myth or a modern 'urban legend' whizzing around the internet, it is also worth stopping to ask whether it – or any part of it – is true. So let's ask ourselves that question – Who was the first person? – and take a look at the true, scientific answer.

Who was the first person *really?*

THIS MAY surprise you, but there never was a first person – because every person had to have parents, and those parents had to be people too! Same with rabbits. There never was a first rabbit, never was a first crocodile, never a first dragonfly. Every creature ever born belonged to the same species as its parents (with perhaps a very small number of exceptions, which I shall ignore here). So that must mean that every creature ever born belonged to the same species as its grandparents. And its great-grandparents. And its great-great-grandparents. And so on for ever.

For ever? Well, no, it's not as simple as that. This is going to need a bit of explaining, and

I'll begin with a thought experiment. A thought experiment is an experiment in your imagination. What we are going to imagine is not literally possible because it takes us way, way back in time, long before we were born. But *imagining* it teaches us something important. So, here is our thought experiment. All you have to do is imagine yourself following these instructions.

Find a picture of yourself. Now take a picture of your father and place it on top. Then find a picture of his father, your grandfather. Then place on top of that a picture of your grandfather's father, your great-grandfather.

You may not have ever met any of your great-grandfathers. I never met any of mine, but I know that one was a country schoolmaster, one a country doctor, one a forester in British India, and one a lawyer, greedy for cream, who died rock-climbing in old age. Still, even if you don't know what your father's father's father looked like, you can imagine him as a sort of shadowy figure, perhaps a fading brown photograph in a leather frame. Now do the same thing with his father, your great-great-grandfather. And just carry on piling the pictures on top of each other, going back through more and more and more great-great-greats. You can go on doing this even before photography was invented: this is a *thought* experiment, after all.

How many greats do we need for our thought experiment? Oh, a mere 185 million or so will do nicely!

Mere?

MERE?

It isn't easy to imagine a pile of 185 million pictures. How high would it be? Well, if each picture was printed as a normal picture postcard, 185 million pictures would form a tower about 220,000 feet high: that's more than 180 New York skyscrapers standing on top of each other. Too tall to climb, even if it didn't fall over (which it would). So let's tip it safely on its side, and pack the pictures along the length of a single bookshelf.

How long is the bookshelf?
About forty miles.

The near end of the bookshelf has the picture of you. The far end has a picture of your 185-million-greats-grandfather. What did he look like? An old man with wispy hair and white sidewhiskers? A caveman in a leopard skin? Forget any such thought. We don't know exactly what he looked like, but fossils give us a pretty good idea. Your 185-million-greats-grandfather looked something like this ————————➤

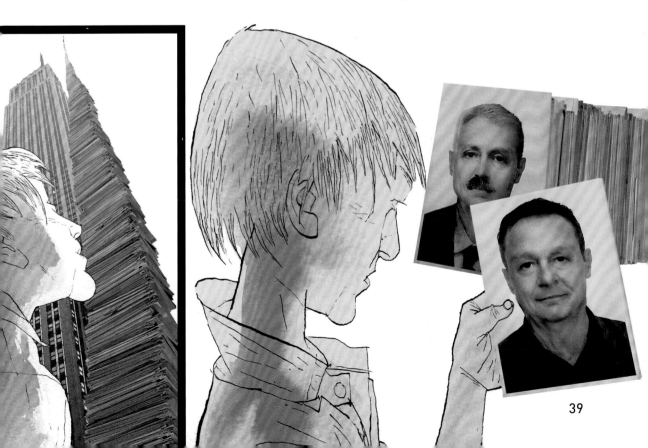

a fish at the other. And lots of other interesting great-...great-grandparents in between, which, as we shall soon see, include some animals that look like apes, others that look like monkeys, others that look like shrews, and so on. Each one is like its neighbours in the line, yet if you pick any two pictures far apart in the line they are very different – and if you follow the line from humans back far enough you come to a fish. How can this be?

Actually, it isn't all that difficult to understand. We are quite used to gradual changes that, step by tiny step, one after the other, make up a big change. You were once a baby. Now you are not. When you are a lot older you'll look quite different again. Yet every day of your life, when you wake up, you are the same person as when you went to bed the previous night. A baby changes into a toddler, then into a child, then into an adolescent; then a young adult, then a middle-aged adult, then an old person. And the change happens so gradually that there never is a day when you can say, 'This person has suddenly stopped being a baby and become a toddler.' And later on there never comes a day when you can say, 'This person has stopped being a child and become an adolescent.' There's never a day when you can say, 'Yesterday this man was middle-aged: today he is old.'

That helps us to understand our thought experiment, which takes us back through 185 million generations of parents and grandparents and great-grandparents until we come face to face with a fish. And, turning round to go forwards in time, it's

Yes, that's right. Your 185-million-greats-grandfather was a fish. So was your 185-million-greats-grandmother, which is just as well or they couldn't have mated with each other and you wouldn't be here.

Let's now walk along our forty-mile bookshelf, pulling pictures off it one by one to have a look at them. Every picture shows a creature belonging to the same species as the picture on either side of it. Every one looks just like its neighbours in the line – or at least as much alike as any man looks like his father and his son. Yet if you walk steadily from one end of the bookshelf to the other, you'll see a human at one end and what happened when your fish ancestor had a fishy child, who had a fishy child, who had a child . . . who, 185 million (gradually less fishy) generations later, turned out to be you.

So it was all very gradual – so gradual that you wouldn't notice any change as you walked back a thousand years; or even ten thousand years, which would bring you to somewhere around your 400-greats-grandfather. Or rather, you would notice lots of little changes all the way along, because nobody looks exactly like their father. But you wouldn't notice any general *trend*. Ten thousand years back from modern humans is not long enough to show a trend. The portrait of your ancestor of ten thousand years ago would be no different from modern people, if we set aside superficial differences in dress and hair and whisker style. He would be no more different from us than modern people are different from other modern people.

How about a hundred thousand years, where we might find your 4,000-greats-grandfather? Well, now, maybe there would be a just-noticeable change. Perhaps a slight thickening of the skull, especially under the eyebrows. But it would still only be slight. Now let's push a bit further back in time. If you walked the first million years along the shelf, the picture of your 50,000-greats-grandfather would be different enough to count as a different species, the one we call *Homo erectus*. We today, as you know, are *Homo sapiens*. *Homo erectus* and *Homo sapiens* probably wouldn't have been able to mate with each other; or, even if they could, the baby would probably not have been able to have babies of its own – in the same way that a mule, which has a donkey father and a horse mother, is almost always unable to have offspring. (We'll see why in the next chapter.)

Once again, though, everything is gradual. You are *Homo sapiens* and your 50,000-greats-grandfather was *Homo erectus*. But there never was a *Homo erectus* who suddenly gave birth to a *Homo sapiens* baby.

So, the question of who was the first person, and when they lived, doesn't have a precise answer. It's kind of fuzzy, like the answer to the question: When did you stop being a baby and become a toddler? At some point, probably less than a million years ago but more than a hundred thousand years ago, our ancestors were sufficiently different from us that a modern person wouldn't have been able to breed with them if they had met.

Whether we should call *Homo erectus* a person, a human, is a different question. That's

Your 50,000-greats-grandfather

Your 4,000-greats-grandfather

a question about how you choose to use words – what's called a semantic question. Some people might want to call a zebra a stripy horse, but others might like to keep the word 'horse' for the species that we ride. That's another semantic question. You might prefer to keep the words 'person', 'man' and 'woman' for *Homo sapiens*. That's up to you. Nobody, however, would want to call your fishy 185-million-greats-grandfather a man. That would just be silly, even though there is a continuous chain linking him to you, every link in the chain being a member of exactly the same species as its neighbours in the chain.

Turned to stone

Now, how do we know what our distant ancestors looked like, and how do we know when they lived? Mostly from fossils. All the pictures of our ancestors in this chapter are reconstructions based on fossils but coloured by comparing them with modern animals.

Fossils are made of stone. They are stones that have picked up the shapes of dead animals or plants. The great majority of animals die with no hope of turning into a fossil. The trick, if you want to be a fossil, is to get yourself buried in the right kind of mud or silt, the kind that might eventually harden to form 'sedimentary rock'.

What does that mean? Rocks are of three kinds: igneous, sedimentary and metamorphic. I shall ignore metamorphic rocks, as they were originally one of the other two kinds, igneous or sedimentary, and have been changed by pressure and/or heat. Igneous rocks (from the Latin for

'fire', *ignis*) were once molten, like the hot lava that comes out of erupting volcanoes now, and solidified into hard rock when they cooled. Hard rocks, of any kind, get worn down ('eroded') by wind or water to make smaller rocks, pebbles, sand and dust. Sand or dust gets suspended in water and can then settle in layers of *sediment* or mud at the bottom of a sea, lake or river. Over a very long time, sediments can harden to make layers (or 'strata') of *sedimentary rock*. Although all strata start off flat and horizontal, they have often got tilted, upended or warped by the time we see them, millions of years later (for how this happens, see Chapter 10 on earthquakes).

Now, suppose a dead animal happens to get washed into the mud, in an estuary perhaps. If the mud later hardens to become sedimentary rock, the animal's body may rot away, leaving in the hardening rock a hollow imprint of its form which we eventually find. That is one kind of fossil – a kind of 'negative' picture of the animal. Or the hollow imprint may act as a mould into which new sediments fall, later hardening to form a 'positive' replica of the outside of the animal's body. That's a second kind of fossil. And there's a third kind of fossil in which the atoms and molecules of the animal's body are, one by one, replaced by atoms and molecules of minerals from the water, which later crystallize to form rock. This is the best kind of fossil because, with luck, tiny details of the animal's insides are permanently reproduced, right through the middle of the fossil.

Fossils can even be dated. We can tell how old they are, mostly by measuring radioactive isotopes in the rocks. We'll learn what isotopes are, and atoms, in Chapter 4. Briefly, a radioactive isotope is a kind of atom which decays into a different kind of atom: for example, one called uranium-238 turns into one called lead-206. Because we know how long this takes to happen, we can think of the isotope as a radioactive clock. Radioactive clocks are rather like the water clocks and candle clocks that people used in the days before pendulum clocks were invented. A tank of water with a hole in the bottom will drain at a measurable rate. If the tank was filled at dawn, you can tell how much of the day has passed by measuring the present level of water. Same with a candle clock. The candle burns at a fixed rate, so you can tell how long it has been burning by measuring how much candle is left. In the case of a uranium-238 clock, we know that it takes 4.5 billion years for half the uranium-238 to decay to lead-206. This is called the 'half-life' of uranium-238. So, by measuring how much lead-206 there is in a rock, compared with the amount of uranium-238, you can calculate how long it is since there was no lead-206 and only uranium-238: how long, in other words, since the clock was 'zeroed'.

And when is the clock zeroed? Well, it only happens with igneous rocks, whose clocks are all zeroed at the moment when the molten rock hardens to become solid. It doesn't work with sedimentary rock, which has no such 'zero moment', and this is a pity because fossils are found only in sedimentary rocks. So we have to find igneous rocks close by sedimentary layers and use them as our clocks. For example, if a fossil is in a sediment with 120-million-year-old igneous rock above it and 130-million-year-old igneous rock below it, you know the fossil dates from somewhere between 120 million and 130 million years ago. That's how all the dates

I mention in this chapter are arrived at. They are all approximate dates, not to be taken as too precise.

Uranium-238 is not the only radioactive isotope we can use as a clock. There are plenty of others, with a wonderfully wide spread of half-lives. For example, carbon-14 has a half-life of only 5,730 years, which makes it useful for archaeologists looking at human history. It is a beautiful fact that many of the different radioactive clocks have overlapping timescales, so we can use them to check up on each other. And they always agree.

The carbon-14 clock works in a different way from the others. It doesn't involve igneous rocks but uses the remains of living bodies themselves, for example old wood. It is one of the fastest of our radioactive clocks, but 5,730 years is still much longer than a human lifetime, so you might ask how we know it is the half-life of carbon-14, let alone how we know that 4.5 billion years is the half-life of uranium-238! The answer is easy. We don't have to wait for half of the atoms to decay. We can measure the rate of decay of only a tiny fraction of the atoms, and work out the half-life (quarter-life, hundredth-life, etc.) from that.

A ride back in time

Let's do another thought experiment. Take a few companions and get in a time machine. Fire up the engine and zoom back ten thousand years. Open the door and have a look at the people you meet. If you happen to land in what is now Iraq,

they'll be in the process of inventing agriculture. In most other places they'll be 'hunter-gatherers', moving from place to place, hunting wild animals and gathering wild berries, nuts and roots. You won't be able to understand what they say and they will be wearing very different clothes (if any). Nevertheless, if you dress them in modern clothes and give them modern haircuts, they will be indistinguishable from modern people (or no more different from some modern people than people are different from one another today). And they will be fully capable of breeding with any of the modern people on board your time machine.

Now, take one volunteer from among them (perhaps your 400-greats-grandfather, because this is approximately the time when he might have lived) and set off again in your time machine, back another ten thousand years: to twenty thousand years ago, where you have a chance to meet your 800-greats-grandparents. This time the people you see will all be hunter-gatherers but, once again, their bodies will be those of fully modern humans and, once again, they will be perfectly capable of interbreeding with modern people and producing fertile offspring. Take one of them with you in the time machine, and set off another ten thousand years into the past. Keep on doing this, hopping back in steps of ten thousand years, at each stop picking up a new passenger and taking him or her back to the past.

The point is that eventually, after a lot of ten-thousand-year hops, perhaps when you've gone a

million years into the past, you'll begin to notice that the people you meet when you emerge from the time machine are definitely different from us, and can't interbreed with those of us who boarded with you at the start of its journey. But they will be capable of breeding with the latest additions to the passenger list, who are almost as ancient as they are themselves.

I'm just making the same point as I made before – about gradual change being imperceptible, like the moving hour hand of a watch – but using a different thought experiment. It's worth saying in two different ways, because it is so important and yet – quite understandably – so hard for some people to appreciate.

Let's resume our journey into the past, and look at some of the stations on the way back to that beautiful fish. Suppose we have just arrived in our time machine at the station labelled 'Six Million Years Ago'. What shall we find there? So long as we make a point of being in Africa, we'll find our 250,000-greats-grandparents (give or take some generations). They'll be apes, and they might look a bit like chimpanzees. But they won't be chimpanzees. Instead, they'll be the ancestors that we share with chimpanzees. They'll be too different from us to mate with us, and too different from chimpanzees to mate with chimpanzees. But they will be able to mate with the passengers we took on board at Station Five Million Nine Hundred and Ninety Thousand Years Ago. And probably those from Station Five Million Nine Hundred Thousand Years Ago, too. But probably not those who joined us at Station Four Million Years Ago.

Let's now resume our ten-thousand-year hops, all the way back to Station Twenty-Five Million Years Ago. There we shall find your (and

Your 250,000-greats-grandfather
(6 million years ago)

46

my) one-and-a-half-million-greats-grandparents – at an approximate estimate. They will not be apes, for they will have tails. We would call them monkeys if we met them today, although they are no more closely related to modern monkeys than they are to us. Although very different from us, and incapable of breeding with us or with modern monkeys, they will breed happily with the all-but-identical passengers who joined us at Station Twenty-Four Million Nine Hundred and Ninety Thousand Years Ago. Gradual, gradual change, all the way.

On we go, back and back, ten thousand years at a time, finding no noticeable change at each stop. Let's pause to see who greets us when we reach Station Sixty-Three Million Years Ago. Here we can shake hands (paws?) with our seven-million-greats-grand-parents. They look something like lemurs or bushbabies, and they are indeed the ancestors of all modern lemurs and bushbabies, as well as the ancestors of all modern monkeys and apes, including us.

Your 1,500,000-greats-grandfather
(25 million years ago)

They are as closely related to modern humans as they are to modern monkeys, and no more closely to modern lemurs or bushbabies. They wouldn't be able to mate with any modern animals. But they would probably be able to mate with the passengers we picked up at Station Sixty-Two Million Nine Hundred and Ninety Thousand Years Ago. Let's welcome them aboard the time machine, and speed on backwards.

Your 7,000,000-greats-grandfather
(63 million years ago)

Your 45,000,000-greats-grandfather
(105 million years ago)

At Station One Hundred and Five Million Years Ago we'll meet our 45-million-greats-grandfather. He is also the grand ancestor of all the modern mammals except marsupials (now found mostly in Australia, plus a few in America) and monotremes (duckbilled platypuses and spiny anteaters, now found only in Australia/New Guinea). The picture shows him with his favourite food, an insect, in his mouth. He is equally closely related to all modern mammals, although he may look a bit more like some of them than others.

Station Three Hundred and Ten Million Years Ago presents us with our 170-million-greats-grandmother. She is the grand ancestor of all modern mammals, all modern reptiles – snakes, lizards, turtles, crocodiles – and all dinosaurs (including birds, because birds arose from within the dinosaurs). She is equally distantly related to all those modern animals, although she looks more like a lizard. What that means is that lizards have changed less since her time than, say, mammals have.

Seasoned time-travellers as we are by now, it isn't far to go until we hit the fish that I mentioned earlier. Let's make one more stop on the way: at Station Three

Hundred and Forty Million Years Ago, where we meet our 175-million-greats-grandfather. He looks a bit like a newt, and is the grand ancestor of all modern amphibians (newts and frogs) as well as of all the other land vertebrates.

And so to Station Four Hundred and Seventeen Million Years Ago and your 185-million-greats-grandfather, the fish on page 40. From there we could go on even further back in time, meeting more and more distant great-grandparents, including various kinds of fish with jaws, then fish without jaws, then . . . well, then our knowledge starts to fade into a kind of mist of uncertainty, for these very ancient times are where we start to run out of fossils.

Your 175,000,000 -
greats-grandfather
(340 million years ago)

Your 170,000,000 -
greats-grandmother
(310 million years ago)

49

DNA tells us we are all cousins

Although we may lack the fossils to tell us exactly what our very ancient ancestors looked like, we are in no doubt at all that all living creatures are our cousins, and cousins of each other. And we also know which modern animals are close cousins of each other (like humans and chimpanzees, or rats and mice), and which are distant cousins of each other (like humans and cuckoos, or mice and alligators). How do we know? By systematically comparing them. Nowadays, the most powerful evidence comes from comparing their DNA.

DNA is the genetic information that all living creatures carry in each of their cells. The DNA is spelled out along massively coiled 'tapes' of data, called 'chromosomes'. These chromosomes really are very like the kind of data tapes you'd feed into an old-fashioned computer, because the information they carry is *digital* and is strung along them in order. They consist of long strings of code 'letters', which you can count: each letter is either there or it isn't – there are no half measures. That's what makes it digital, and why I say DNA is 'spelled out'.

All genes, in every animal, plant and bacterium that has ever been looked at, are coded messages for how to build the creature, written in a standard alphabet. The alphabet has only four letters to choose from (as opposed to the 26 letters of the English alphabet), which we write as A, T, C and G. The same genes occur in many different creatures, with a few revealing differences. For example, there's a gene called FoxP2, which is shared by all mammals and lots more creatures besides. The gene is a string of more than 2,000 letters. At the bottom of this page is a short stretch of 80 letters from somewhere in the middle of FoxP2, the stretch from letter number 831 to letter number 910. The upper row is from a human, the middle row from a chimpanzee and the bottom row from a mouse. The numbers at the end of the bottom two rows show how many letters in the whole gene are different from those in the whole human FoxP2 gene.

You can tell that FoxP2 is the same gene in all mammals because the great majority of the code letters are the same, and that is true of the whole length of the gene, not just this stretch of 80 letters. Not quite all the chimpanzee letters are the same as ours, and somewhat fewer of the mouse ones are. The differences are highlighted in red. Of the total of 2,076 letters in FoxP2, the chimpanzee has nine letters different from ours, while the mouse has 139 letters different. And that pattern holds for other genes

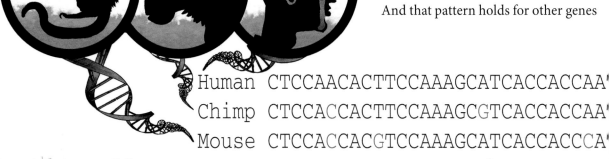

Human CTCCAACACTTCCAAAGCATCACCACCAA
Chimp CTCCACCACTTCCAAAGCGTCACCACCAA
Mouse CTCCACCACGTCCAAAGCATCACCACCCA

too. That explains why chimpanzees are very like us, while mice are less so.

Chimpanzees are our close cousins, mice are our more distant cousins. 'Distant cousins' means that the most recent ancestor we share with them lived a long time ago. Monkeys are closer to us than mice but further from us than chimpanzees. Baboons and rhesus macaques are both monkeys, close cousins of each other, and with almost identical FoxP2 genes. They are exactly as distant from chimps as they are from us; and the number of DNA letters in FoxP2 that separate baboons from chimps is almost exactly the same (24) as the number of letters that separate baboons from us (23). It all fits.

And, just to finish off this little thought, frogs are much more distant cousins of all mammals. All mammals have approximately the same number of letter differences from a frog, for the simple reason that they are all *exactly* equally close cousins: all mammals share a more recent ancestor with each other (about 180 million years ago) than they do with the frog (about 340 million years ago).

But of course not all humans are the same as all other humans, and not all baboons are the same as all other baboons and not all mice are the same as all other mice. We could compare your genes with mine, letter by letter. And the result? We'd turn out to have even more letters in common than either of us does with a chimpanzee. But we'd still find some letters that are different. Not many, and there's no particular reason to single out the FoxP2 gene. But if you counted up the number of letters all humans share in all our genes, it would be more than any of us shares with a chimpanzee. And you share more letters with your cousin than you share with me. And you share even more letters with your mother and your father, and (if you have one) with your sister or brother. In fact, you can work out how closely related any two people are to each other by counting the number of DNA letters they share. It's an interesting count to make, and it is something we are probably going to hear more about in the future. For example, the police will be able to track somebody down if they have the DNA 'fingerprint' of his brother.

Some genes are recognizably the same (with minor differences) in all mammals. Counting the number of letter differences in such genes is useful for working out how closely related different mammal species are. Other genes are useful for working out more distant relationships, for example between vertebrates and worms. Other genes again are useful for working out relationships within a species – say, for working out how closely related you are to me. In case you are interested, if you happen to come from England, our most recent shared ancestor probably lived only a few centuries back. If you happen to be a native Tasmanian or a native American we'd have to go back some tens of thousands of years to find a shared ancestor. If you happen to be a !Kung San of the Kalahari Desert, we might have to go back even further.

TCATTCCATAGTGAATGGACAGTCTTCAGTTCTAAGTGCAAGAC

TCATTCCATCGTGAATGGACAGTCTTCAGTTCTAAATGCAAGAC 9

TCATTCCATAGTGAACGGACAGTCTTCAGTTCTGAATGCAAGGC 139

What is a fact beyond all doubt is that we share an ancestor with every other species of animal and plant on the planet. We know this because some genes are recognizably the same genes in all living creatures, including animals, plants and bacteria. And, above all, the genetic code itself – the dictionary by which all genes are translated – is the same across all living creatures that have ever been looked at. We are all cousins. Your family tree includes not just obvious cousins like chimpanzees and monkeys but also mice, buffaloes, iguanas, wallabies, snails, dandelions, golden eagles, mushrooms, whales, wombats and bacteria. All are our cousins. Every last one of them. Isn't that a far more wonderful thought than any myth? And the most wonderful thing of all is that we know for certain it is literally true.

Your 185,000,000 - greats-grandfather
(417 million years ago)

THERE ARE LOTS of myths that attempt to explain why particular kinds of animals are the way that they are – myths that 'explain' things like why leopards have spots, and why rabbits have white tails. But there don't seem to be many myths about the sheer range and variety of different kinds of animals. I can find nothing akin to the Jewish myth of the Tower of Babel, which accounts for the great variety of languages. Once upon a time, according to this myth, all the people in the world spoke the same language. They could there-fore work harmoniously together to build a great tower, which they hoped would reach the sky. God noticed this and took a very dim view of everybody being able to under-stand everybody else. Whatever might they get up to next, if they could talk to each other and work

MANY *different kinds of* animals?

together? So he decided to 'confound their language' so that 'they may not understand one another's speech'. This, the myth tells us, is why there are so many different languages, and why, when people try to talk to people from another tribe or country, their speech often sounds like meaningless babble. Oddly enough, there is no connection between the word 'babble' and the Tower of Babel.

I was hoping to find a similar myth about the great diversity of animals, because there is a resemblance between language evolution and animal evolution, as we shall see. But there doesn't seem to be any myth that specifically tackles the sheer *number* of *different kinds* of animals. This is surprising, because there is indirect evidence that tribal peoples can be well aware of the fact there are many different kinds of animals. In the 1920s a now famous German scientist called Ernst Mayr did a pioneering study of the birds of the New Guinea highlands. He compiled a list of 137 species, then discovered, to his amazement, that the local Papuan tribesmen had separate names for 136 of them.

Back to the myths. The Hopi tribe of North America had a goddess called Spider Woman.

In their creation myth she teamed up with Tawa the sun god, and they sang the First Magic Song as a duet. This song brought the Earth, and life, into being. Spider Woman then took the threads of Tawa's thoughts and wove them into solid form, creating fish, birds, and all other animals.

Other North American tribes, the Pueblo and Navajo peoples, have a myth of life that is a tiny bit like the idea of evolution: life emerges from the Earth like a sprouting plant growing up through a sequence of stages. The insects climbed from their world, the First or Red World, up into the Second World, the Blue World, where the birds lived. The Second World then became too crowded, so the birds and insects flew up into the Third or Yellow World, where the people and other mammals lived. The Yellow World in turn became crowded and food became scarce, so they all, insects, birds and everybody, went up to the Fourth World, the Black and White World of day and night. Here the gods

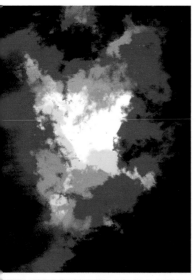

had already created cleverer people who knew how to farm the Fourth World and who taught the newcomers how to do it too.

The Jewish creation myth comes closer to doing justice to diversity, but it doesn't really attempt to explain it. Actually, the Jewish holy book has two different creation myths, as we saw in the previous chapter. In the first one, the Jewish god created everything in six days. On the fifth day he created fish, whales and all sea creatures, and the birds of the air. On the sixth day he made the rest of the land animals, including man. The language of the myth pays some attention to the number and variety of living creatures – for example, 'God created great whales, and every living creature that moveth, which the waters brought forth abundantly after their kind, and every winged fowl after his kind,' and made every 'beast of the earth' and 'every thing that creepeth upon the earth after his kind'. But why was there such variety? We are not told.

In the second myth we get some hint that the god might have thought his first man needed a variety of companions. Adam, the first man, is created alone and placed in the beautiful oasis garden. But then the god realized that 'It is not good that the man should be alone' and he therefore 'formed every beast of the field and every fowl of the air; and brought them unto Adam to see what he would call them'.

57

Why are there REALLY so many different kinds of animals?

PANTHERA L

ADAM'S TASK of naming all the animals was a tough one – tougher than the ancient Hebrews could possibly have realized. It's been estimated that about 2 million species have so far been given scientific names, and even these are just a small fraction of the number of species yet to be named.

How do we even decide whether two animals belong in the same species or in two different species? Where animals reproduce sexually, we can come up with a sort of definition. Animals belong to different species if they don't breed together. There are borderline cases like horses and donkeys, which can breed together but produce offspring (called mules or hinnies) that are infertile – that is, that cannot have offspring themselves. We therefore place a horse and

a donkey in different species. More obviously, horses and dogs belong to different species because they don't even try to interbreed, and couldn't produce offspring if they did, even infertile ones. But spaniels and poodles belong to the same species because they happily interbreed, and the puppies that they produce are fertile.

Every scientific name of an animal or plant consists of two Latin words, usually printed in *italics*. The first word refers to the 'genus' or group of species and the second to the individual species within the genus. *Homo sapiens* ('wise man') and *Elephas maximus* ('very big elephant') are examples. Every species is a member of a genus. *Homo* is a genus. So is *Elephas*. The lion is *Panthera leo* and the genus *Panthera* also

CARNIVORA

FELIDAE

THERA ONCA

includes *Panthera tigris* (tiger), *Panthera pardus* (leopard or 'panther') and *Panthera onca* (jaguar). *Homo sapiens* is the only surviving species of our genus, but fossils have been given names like *Homo erectus* and *Homo habilis*. Other human-like fossils are sufficiently different from *Homo* to be placed in a different genus, for example *Australopithecus africanus* and *Australopithecus afarensis* (nothing to do with Australia, by the way: australo- just means 'southern', which is where Australia's name also comes from).

Each genus belongs to a *family*, usually printed in ordinary 'roman' type with a capital initial. Cats (including lions, leopards, cheetahs, lynxes and lots of smaller cats) make up the family Felidae. Every family belongs to an *order*. Cats, dogs, bears, weasels and hyenas belong to different families within the order Carnivora. Monkeys, apes (including us) and lemurs all belong to different families within the order Primates. And every order belongs to a class. All mammals are in the class Mammalia.

Can you see the shape of a tree developing in your mind as you read this description of the sequence of groupings? It is a family tree: a tree with many branches, each branch having sub-branches, and each sub-branch having sub-sub-branches. The tips of the twigs are species. The other groupings – class, order, family, genus – are the branches and sub-branches. The whole tree is all of life on Earth.

Think about why trees have so many twigs. Branches branch. When we have enough branches of branches of branches, the total number of twigs can be very large. That's what happens in evolution. Charles Darwin himself drew a branching tree as the only picture in his most famous book, *On the Origin of Species*. Below is an early version of Darwin's tree picture, which he sketched in one of his notebooks some years earlier. At the top of the page he wrote a mysterious little message to himself: 'I think'. What do you think he meant? Maybe he started to write a sentence and one of his children interrupted him so he never finished it. Maybe he found it easier to represent quickly what he was thinking in this diagram than in words. Perhaps we shall never know. There is other handwriting on the page, but it is hard to decipher. It is tantalizing to read the actual notes of a great scientist, written on a particular day and never meant for publication.

The following isn't exactly how the tree of animals branched, but it gives you an idea of the principle. Imagine an ancestral species splitting into two species. If each of those then splits into two, that makes four. If each of them splits into two, that makes eight, and so on through 16, 32, 64, 128, 256, 512 . . . You can see that, if you carry on doubling up, it doesn't take long to get up into the millions of species. That probably makes sense to you, but you may be wondering why a species should split. Well, it's for pretty much the same reason as human languages split, so let's pause to think about that for a moment.

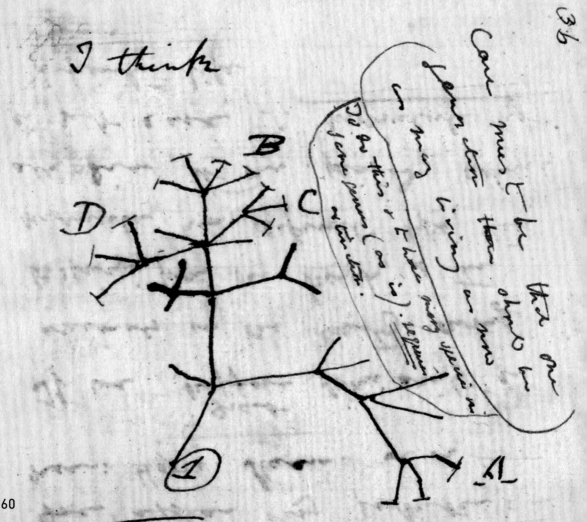

Pulling apart: how languages, and species, divide

Although the legend of the Tower of Babel is, of course, not really true, it does raise the interesting question of why there are so many different languages.

Just as some species are more similar than others and are placed in the same family, so there are also families of languages. Spanish, Italian, Portuguese, French and many European languages and dialects such as Romansch, Galician, Occitan and Catalan are all pretty similar to each other; together they're called 'Romance' languages. The name actually comes from their common origin in Latin, the language of Rome, not from any association with romance, but let's use an expression of love as our example. Depending on which country you are in, you might declare your feelings in one of the following ways: 'Ti amo', 'Amote', 'T'aimi' or 'Je t'aime'. In Latin it would be 'Te amo' – exactly like modern Spanish.

To swear your love to someone in Kenya, Tanzania or Uganda you could say, in Swahili, 'Nakupenda'. A bit further south, in Mozambique, Zambia, or Malawi where I was brought up, you might say, in the Chinyanja language, 'Ndimakukonda'. In other so-called Bantu languages in southern Africa you might say 'Ndinokuda', 'Ndiyakuthanda' or, to a Zulu, 'Ngiyakuthanda'. This Bantu family of languages is quite distinct from the Romance family of languages, and both are distinct from the Germanic family which includes Dutch, German and the Scandinavian languages. See how we use the word 'family' for languages, just as we do for species (the cat family, the dog family) and also, of course, for our own families (the Jones family, the Robinson family, the Dawkins family).

It isn't hard to work out how families of related languages arise over the centuries. Listen

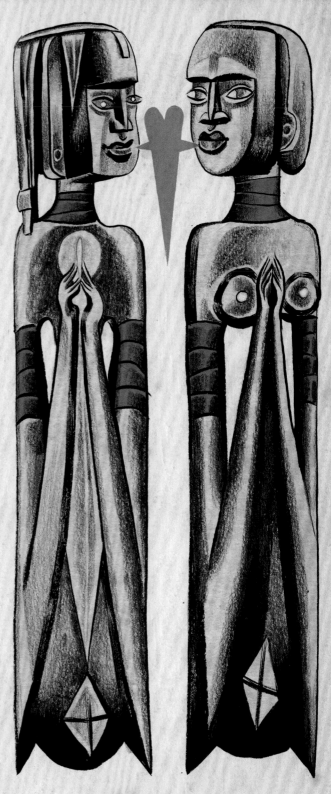

to the way you and your friends speak to each other, and compare it to the way your grandparents speak. Their speech is only slightly different and you can easily understand them, but they are only two generations away. Now imagine talking, not to your grandparents but to your 25-greats-grandparents. If you happen to be English, that might take you back to the late fourteenth century – the lifetime of the poet Geoffrey Chaucer, who wrote descriptions like this:

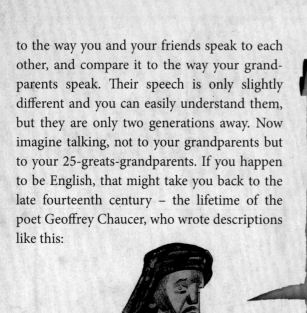

He was a lord ful fat and in good poynt;
His eyen stepe, and rollynge in his heed,
That stemed as a forneys of a leed;
His bootes souple, his hors in greet estaat.
Now certeinly he was a fair prelaat;
He was nat pale as a forpyned goost.
A fat swan loved he best of any roost.
His palfrey was as broun as is a berye.

Well, it is recognizably English, isn't it? But I bet you'd have a hard time understanding it if you heard it spoken. (If you'd like to try, you can listen to me reading Chaucer here: **www.booksattransworld. co.uk/dawkins-chaucer**) And if it was any more different you'd probably consider it a separate language, as different as Spanish is from Italian.

So, the language in any one place changes century by century. We could say it 'drifts' into something different. Now add the fact that people speaking the same language in different places don't often have the opportunity to hear each other (or at least they didn't before telephones and radios were invented); and the fact that language drifts in different directions in different places. This applies to the way it is spoken as well as to the words themselves: think how different English sounds in a Scottish, Welsh, Geordie, Cornish, Australian or American accent. And Scottish people can easily distinguish an Edinburgh accent from a Glasgow accent or a Hebridean accent. Over time, both the way the language is spoken and the words used become characteristic of a region; when two ways of speaking a language have drifted sufficiently far apart, we call them different 'dialects'.

After enough centuries of drift, different regional dialects eventually become so different that people in one region can no longer understand people in another. At this point we call them separate languages. That is what happened when German and Dutch drifted, in separate directions, from a now extinct ancestral language. It is what happened when French, Italian, Spanish and Portuguese independently drifted away from Latin in separate parts of Europe.

You can draw a family tree of languages, with 'cousins' like French, Portuguese and Italian on neighbouring 'branches' and ancestors like Latin further down the tree – just as Darwin did with species.

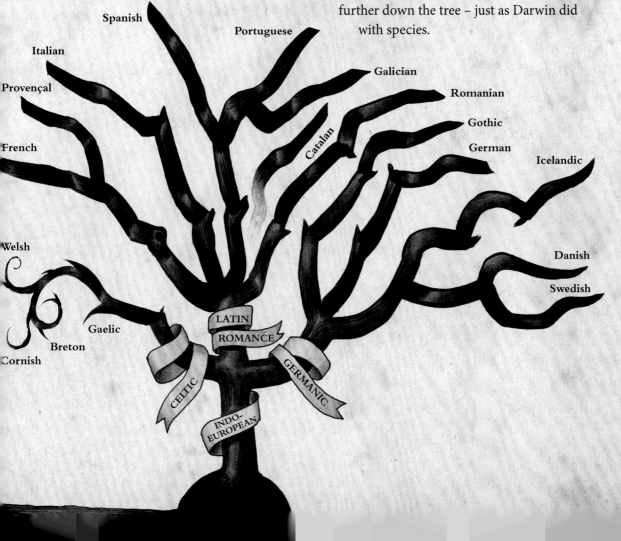

Like languages, species change over time and over distance. Before we look at *why* this happens, we need to see *how* they do it. For species, the equivalent of words is DNA – the genetic information every living thing carries inside it that determines how it is made, as we saw in Chapter 2. When individuals reproduce sexually, they mix their DNA. And when members of one local population migrate into another local population and introduce their genes into it by mating with individuals of the population they have just joined, we call this 'gene flow'.

The equivalent of, say, Italian and French drifting apart is that the DNA of two separated populations of a species becomes less and less alike over time. Their DNA becomes less and less able to work together to make babies. Horses and donkeys can mate with each other, but horse DNA has drifted so far from donkey DNA that the two can no longer understand each other. Or

rather, they can mix well enough – the two 'DNA dialects' can understand each other well enough – to make a living creature, a mule, but not well enough to make one that can reproduce itself: mules, as we saw earlier, are sterile.

An important difference between species and languages is that languages can pick up 'loan words' from other languages. Long after it developed as a separate language from Romance, Germanic and Celtic sources, for example, English picked up 'shampoo' from Hindi, 'iceberg' from Norwegian, 'bungalow' from Bengali and 'anorak' from Inuit. Animal species, by contrast, never (or almost never) exchange DNA ever again, once they have drifted far enough apart to have stopped breeding together. Bacteria are another story: they do exchange genes, but there isn't enough space in this book to go into that. In the rest of this chapter, assume that I am talking about animals.

Islands and isolation: the power of separation

So the DNA of species, like the words of languages, drifts apart when separated. Why might this happen? What might start the separation? An obvious possibility is the sea. Populations on separate islands don't meet each other – not often, anyway – so their two sets of genes have the opportunity to drift away from one another. This makes islands extremely important in the origins of new species. But we can think of an island as more than just a piece of land surrounded by water. To a frog, an oasis is an 'island' where it can live, surrounded by a desert where it can't. To a fish, a lake is an island. Islands matter, both for species and for languages, because the population of an island is cut off from contact with other populations (preventing gene flow in the case of species, just as it prevents language drift) and so is free to begin to evolve in its own direction.

The next important point is that the population of an island need not be totally isolated for ever: genes can occasionally cross the barrier surrounding it, whether this be water or uninhabitable land.

On 4 October 1995 a mat of logs and uprooted trees was blown onto a beach on the Caribbean island of Anguilla. On the mat were 15 green iguanas, alive after what must have been a perilous journey from another island, probably Guadeloupe, 160 miles away. Two hurricanes, called Luis and Marilyn, had roared through the Caribbean during the previous month, uprooting trees and flinging them into the sea. It seems that one of these hurricanes must have torn down the trees in which the iguanas were climbing (they love sitting up in trees, as I have seen in Panama) and blown them out to sea. Eventually reaching Anguilla, they crawled off their unorthodox means of transport onto the beach and began a new life, feeding and reproducing and passing on their DNA, on a brand new island home.

We know this happened because the iguanas were seen arriving on Anguilla by local fishermen. Centuries earlier, although nobody was there to witness it, something similar is almost certainly what brought the iguanas' ancestors to Guadeloupe in the first place. And something like the same story almost certainly accounts for the presence of iguanas on the Galapagos islands, which is where we turn for the next step in our story.

The Galapagos islands are historically important because they probably inspired Charles Darwin's first thoughts on evolution when, as a

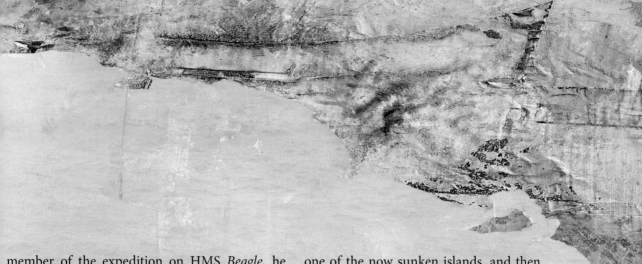

member of the expedition on HMS *Beagle*, he visited them in 1835. They are a collection of volcanic islands in the Pacific Ocean near the equator, about 600 miles west of South America. They are all young (just a few million years old), formed by volcanoes punching up from the bottom of the sea. This means that all the species of animals and plants on the islands must have arrived from elsewhere – presumably the mainland of South America – and recently, by evolutionary standards. Once arrived, species could make the shorter crossings from island to island, sufficiently often to reach all the islands (maybe once or twice every century or so) but sufficiently seldom that they were able to evolve separately – 'drift apart' as we have been saying in this chapter – during the intervals between the rare crossings.

Nobody knows when the first iguanas arrived in the Galapagos. They probably rafted across from the mainland just like the ones that arrived in Anguilla in 1995. Nowadays the nearest island to the mainland is San Cristobal (Darwin knew it by the English name of Chatham), but millions of years ago there were other islands too, which have now sunk beneath the sea. The iguanas could have arrived first on one of the now sunken islands, and then crossed to other islands, including those still above water today.

Once there, they had the opportunity to flourish in a new place, just like the ones that arrived in Anguilla in 1995. The first iguanas on Galapagos would have evolved to become different from their cousins on the mainland, partly by just 'drifting' (like languages) and partly because natural selection would have favoured new survival skills: a relatively barren volcanic island is a very different place from the South American mainland.

The distances between the different islands are much smaller than the distance from any of them to the mainland. So accidental sea crossings between islands would be relatively common: perhaps once per century rather than once per millennium. And iguanas would have started turning up on most or all of the islands eventually. Island-hoppings would have been rare enough to allow some evolutionary drifting apart on the different islands, between 'contaminations' of the genes by subsequent island-hoppings: rare enough to allow them to evolve so much that when they eventually met again they could no longer breed together. The result is that there are now three distinct species of land iguana on Galapagos, which are no longer capable of cross-breeding with each other. *Conolophus pallidus* is found only on the island of Santa Fe. *Conolophus subcristatus* lives on several islands including Fernandina, Isabela and Santa Cruz (each island population possibly on its way to becoming a separate species). *Conolophus marthae* is confined to the northernmost of the chain of five volcanoes on the big island of Isabela.

That raises another interesting point, by the way. You remember I said that a lake or an oasis could count as an island, even though neither consists of land surrounded by water? Well, the same goes for each of the five volcanoes on Isabela. Each volcano in the chain is surrounded by a zone of rich vegetation (green in the picture below), which is a kind of oasis, separated from the next volcano by a desert. Most of the Galapagos islands have only a single large volcano, but Isabela has five. If the sea level rises (perhaps because of global warming) Isabela could become five islands separated by sea. As it is, you can think of each volcano as a kind of island within an island. That's how it would seem to an animal like

a land iguana (or a giant tortoise), which needs to feed on the vegetation found only on the slopes around the volcanoes.

Any kind of isolation by a geographical barrier which can be crossed sometimes but not too often leads to evolutionary branching (Actually, it doesn't have to be a geographical barrier. There are other possibilities, especially in insects, but for simplicity's sake I won't go into them here.) And once the divided populations have drifted far enough apart that they can no longer breed together, the geographical barrier is no longer necessary. The two species can go their separate evolutionary ways without contaminating each other's DNA ever again. It is mainly separations of this kind that were originally responsible for all the new species that have ever arisen on this planet: even, as we shall see, the original separation of the ancestors of, say, snails from the ancestors of all vertebrates including us.

At some point in the history of iguanas on Galapagos, a branching occurred which was to lead to a very peculiar new species. On one of the islands – we don't know which – a local population of land iguanas completely changed their way of life. Instead of eating land plants on the slopes of volcanoes, they went to the shore and took to feeding on seaweed. Natural selection then favoured those individuals that became skilled swimmers, until nowadays their descendants habitually dive to graze on underwater seaweeds. They are called marine iguanas and, unlike land iguanas, they are found nowhere but Galapagos.

They have lots of strange features that equip them for life in the sea and this makes them really rather different from the land iguanas of Galapagos and everywhere else in the world. They have certainly evolved from land iguanas, but they are not especially close cousins of today's land iguanas of Galapagos, so it is possible that they evolved from an earlier, now extinct genus, which colonized the islands from the mainland long before the present *Conolophus*. There are different races of marine iguanas, but not different species, on the different islands. One day these different island races will probably be found to have drifted apart far enough to be called different species of the marine iguana genus.

It's a similar story for giant tortoises, for lava lizards, for the strange flightless cormorants, for mockingbirds, for finches, and for many other animals and plants of Galapagos. And the same kind of thing happens all over the world. Galapagos is just a particularly clear example. Islands (including lakes, oases and mountains) manufacture new species. A river can do the same thing. If it is difficult for an animal to cross a river, the genes in populations on either side of the river can drift apart, just as one language can drift to become two dialects, which can later drift to become two languages. Mountain ranges can play the same role of separation. So can just plain *distance*. Mice in Spain may be connected by a chain of interbreeding mice all across the Asian continent to China. But it takes so long for a gene to travel from mouse to mouse across that vast distance that they might as well be on separate islands. And mouse evolution in Spain and China might drift in different directions.

The three species of Galapagos land iguana have had only a few thousand years to drift apart in their evolution. After enough hundreds of millions of years have passed, the descendants of a single ancestral species can be as different as, say, a cockroach is from a crocodile. In fact it is literally true that once upon a time there was a great-great-great- (lots of greats) grandparent of cockroaches (and lots of other animals including snails and crabs) which was also the grand ancestor (let's use the word 'grancestor') of crocodiles (not to mention all the other vertebrates). But you'd have to go back a very very long way, maybe more than a billion years, before you found a grancestor as grand and ancient as that. That is much too long ago for us even to begin to guess what the original barrier was that separated them in the first place. Whatever it was, it must have been in the sea, because in those far-off days no animals lived on land. Maybe the grancestor species could only live on coral reefs, and two populations found themselves on a pair of coral reefs separated by inhospitable deep water.

As we saw in the previous chapter, you'd only have to go back 6 million years to find the most recent shared grancestor of all humans and chimpanzees. That's recent enough for us to guess at a possible geographical barrier that might have occasioned the original split. It's been suggested that it was the Great Rift Valley in Africa, with humans evolving on the east side and chimpanzees on the west. Later, the chimp ancestral line split into common chimpanzees and pygmy chimpanzees or bonobos: it's been suggested that the barrier in that case was the Congo river. As we saw in the previous chapter, the shared grancestor of all surviving mammals lived about 185 million years ago. Since then, its descendants have branched and branched and branched again, producing all the thousands of species of mammals we see today, including 231 species of carnivores (dogs, cats, weasels, bears etc.), 2000 species of rodents, 88 species of whales and dolphins, 196 species of cloven-hoofed animals (cows, antelopes, pigs, deer, sheep), 16 species in the horse family (horses, zebras, tapirs and rhinos), 87 rabbits and hares, 977 species of bats, 68 species of kangaroos, 18 species of apes (including humans), and lots and lots of species that have gone extinct along the way (including quite a few extinct humans, known only from fossils).

Stirring, selection and survival

I want to round off the chapter by telling the story again in slightly different language. I've already briefly mentioned *gene flow*; scientists also talk of something called the *gene pool*, and I now want to spell out more fully what that means. Of course there can't literally be a pool of genes. The word 'pool' suggests a liquid, in which genes might be stirred around. But genes are found only in the cells of living bodies. So what does it mean to talk of a gene pool?

In every generation, sexual reproduction sees to it that genes are shuffled. You were born with the shuffled genes of your father and your mother, which means the shuffled genes of your four grandparents. The same applies to every individual in the population over the long, long reach of evolutionary time: thousands of years, tens of thousands, hundreds of thousands of years. During that time, this process of sexual shuffling sees to it that the genes within the whole population are so thoroughly shuffled, indeed stirred, that it makes sense to talk of a great, swirling pool of genes: the 'gene pool'.

You remember our definition of a species as a group of animals or plants that can breed with each other? Now you can see why this definition matters. If two animals are members of the same species in the same population, that means their genes are being stirred about in the same gene pool. If two animals are members of

different species they cannot be members of the same gene pool because their DNA cannot mix in sexual reproduction, even if they live in the same country and meet each other frequently. If populations of the same species are geographically separated, their gene pools have the opportunity to drift apart – so far apart, eventually, that if they happen to meet again they can no longer breed together. Now that their gene

72

pools have moved beyond mixing they have become different species and can go on moving further apart for millions of years to the point where they might become as different from one another as humans are from cockroaches.

Evolution means change in a gene pool. Change in a gene pool means that some genes become more numerous, others less. Genes that used to be common become rare, or disappear altogether. Genes that used to be rare become common. And the result is that the shape, or size, or colour, or behaviour of typical members of the species changes: it evolves, because of changes in the numbers of genes in the gene pool. That is what evolution is.

Why should the numbers of different genes change as the generations go by? Well, you might say it would be surprising if they didn't, given such immensities of time. Think of the way language changes over the centuries. Words like 'thee' and 'thou', 'zounds' and 'avast', phrases like 'stap me vitals', have

73

now more or less dropped out of English. On the other hand, the phrase 'I was like' (meaning 'I said'), which would have been incomprehensible as recently as 20 years ago, is now commonplace. So is 'cool' as a term of approval.

So far in this chapter, I haven't needed to go much further than the idea that gene pools in separate populations can drift apart, like languages. But actually, in the case of species, there is much more to it than drifting. This 'much more' is natural selection, the supremely important process that was Charles Darwin's greatest discovery. Even without natural selection, we'd expect gene pools that happen to be separated to drift apart. But they'd drift in a rather aimless fashion. Natural selection nudges evolution in a purposeful direction: namely, the direction of survival. The genes that survive in a gene pool are the genes that are good at surviving. And what makes a gene good at surviving? It helps other genes to build bodies that are good at surviving and reproducing: bodies that survive long enough to pass on the genes that helped them to survive.

Exactly how they do it varies from species to species. Genes survive in bird or bat bodies by helping to build wings. Genes survive in mole bodies by helping to build stout, spade-like hands. Genes survive in lion bodies by helping to build fast-running legs, and sharp claws and teeth. Genes survive in antelope bodies by helping to build fast-running legs, and sharp hearing and eyesight. Genes survive in leaf-insect bodies by making the insects all but indistinguishable from leaves. However different the details, in all species the name of the game is gene survival in gene pools. Next time you see an animal – any animal – or any plant, look at it and say to yourself: what I am looking at is an elaborate machine for passing on the genes that made it. I'm looking at a survival machine for genes.

Next time you look in the mirror,
just think: that is what you are too.

4 What are Things

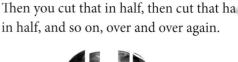

Suppose you take a piece of anything and cut it in half, using the thinnest and sharpest razor blade you can find.

Then you cut that in half, then cut that half in half, and so on, over and over again.

Do the pieces eventually get so small that they can't get any smaller? How thin is the edge of a razor blade? How small is the sharp end of a needle?

IN VICTORIAN TIMES, a favourite book for children was Edward Lear's *Book of Nonsense.* As well as the poems about the Owl and the Pussycat (which you may know because it is still famous), The Jumblies and The Pobble Who Has No Toes, I love the Recipes at the end of the book. The one for Crumboblious Cutlets begins like this:

> *Procure some strips of beef, and having cut them into the smallest possible slices, proceed to cut them still smaller, eight or perhaps nine times.*

What do you get if you keep on cutting stuff into smaller and smaller pieces?

What are the smallest bits that things are made of?

made of?

The ancient civilizations of Greece, China and India all seem to have arrived at the same idea that everything is made from four 'elements': air, water, fire and earth.

But one ancient Greek, Democritus, came a bit closer to the truth. Democritus thought that, if you cut anything up into sufficiently small pieces, you would eventually reach a piece so small that it couldn't be cut any further. The Greek for 'cut' is *tomos*, and if you stick an 'a' in front of a Greek word it means 'not' or 'you can't'. So 'a-tomic' means something too small to be cut any smaller, and that is where our word 'atom' comes from. An atom of gold is the smallest possible bit of gold. Even if it were possible to cut it any smaller, it would cease to be gold. An atom of iron is the smallest possible bit of iron. And so on.

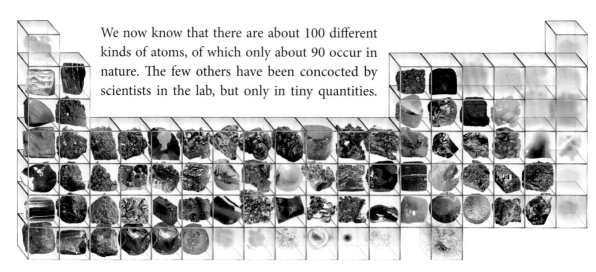

We now know that there are about 100 different kinds of atoms, of which only about 90 occur in nature. The few others have been concocted by scientists in the lab, but only in tiny quantities.

Pure substances that consist of one kind of atom only are called elements (same word as was once used for earth, air, fire and water, but with a very different meaning). Examples of elements are hydrogen, oxygen, iron, chlorine, copper, sodium, gold, carbon, mercury and nitrogen. Some elements, such as molybdenum, are rare on Earth (which is why you may not have heard of molybdenum) but commoner elsewhere in the universe (if you wonder how we know this, wait for Chapter 8).

Metals such as iron, lead, copper, zinc, tin and mercury are elements. So are gases such as oxygen, hydrogen, nitrogen and neon. But most of the substances that we see around us are not elements but compounds. A compound is what you get when two or more different atoms join together in a particular way. You've probably heard water referred to as 'H$_2$O'. This is its chemical formula, and means it is a compound of one oxygen atom joined to two hydrogen atoms. A group of atoms joined together to make a compound is called a *molecule*. Some molecules are very simple: a molecule of water, for example, has just those three atoms. Other molecules, especially those in living bodies, have hundreds of atoms, all joined together in a very particular way. Indeed, it is the way they are joined together, as well as the type and number of atoms, that makes any particular molecule one compound and not another.

You can also use the word 'molecule' to describe what you get when two or more of the same kind of atom join together. A molecule of oxygen, the gas we need in order to breathe, consists of two oxygen atoms joined together. Sometimes three oxygen atoms join together to form a different kind of molecule called ozone. The number of atoms in a molecule really makes a difference, even if the atoms are all the same.

Ozone is harmful to breathe, but we benefit from a layer of it in the Earth's upper atmosphere, which protects us from the most damaging of the sun's rays. One of the reasons Australians have to be especially careful when sunbathing is that there is a 'hole' in the ozone layer in the far south.

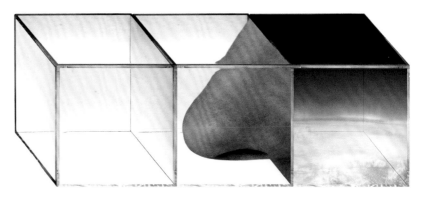

Crystals – atoms on parade

A diamond crystal is a huge molecule, of no fixed size, consisting of millions of atoms of the element carbon stuck together, all lined up in a very particular way. They are so regularly spaced inside the crystal, you could think of them as being like soldiers on parade, except that they are parading in three dimensions, like a shoal of fish. But the number of 'fish' in the shoal – the number of carbon atoms in even the smallest diamond crystal – is gigantic, more than all the fish (plus all the people) in the world. And 'stuck together' is a misleading way to describe them if it makes you think of the atoms as solid lumps of carbon closely packed with no space in between. In fact, as we shall see, most 'solid' matter consists of empty space. That will take some explaining! I'll come back to it.

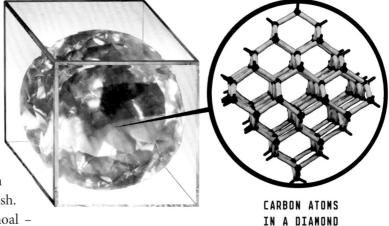

CARBON ATOMS
IN A DIAMOND

All crystals are built up in the same 'soldiers-on-parade' way, with atoms regularly spaced in a fixed pattern that gives the whole crystal its shape. Indeed, that is what we mean by a crystal. Some 'soldiers' are capable of 'parading' in more than one way, producing very different crystals. Carbon atoms, if they parade in one way, make the legendarily hard diamond crystals. But if they adopt a different formation they make crystals of graphite, so soft it is used as a lubricant.

We think of crystals as beautiful transparent objects, and we even describe other things like pure water as 'crystal clear'. But actually, most solid stuff is made of crystals, and most solid stuff is not transparent. A lump of iron is made of lots of tiny crystals packed together, each crystal consisting of millions of iron atoms, spaced out 'on parade' like the carbon atoms in a diamond crystal. Lead, aluminium, gold, copper – all are made of crystals of their different kinds of atoms. So are rocks, like granite or sandstone – but they are often mixtures of lots of different kinds of tiny crystals all packed together.

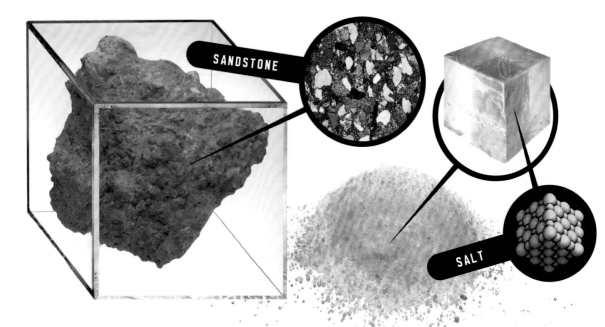

Sand is crystalline, too. In fact, many sand grains are just little bits of rock, ground down by water and wind. The same is true of mud, with the addition of water or other liquids. Often, sand grains and mud grains get packed together again to make new rocks, called 'sedimentary' rocks because they are hardened sediments of sand and mud. (A 'sediment' is the bits of solid stuff that settle in the bottom of a liquid, for example in a river or lake or sea.) The sand in sandstone is mostly made of quartz and feldspar, two common crystals in the Earth's crust. Limestone is different. Like chalk it is calcium carbonate, and it comes from ground-down coral skeletons and sea shells, including the shells of tiny single-celled creatures called forams. If you see a very white beach, the sand is most likely calcium carbonate from the same shelly source.

Sometimes crystals are made entirely of the same kind of atoms 'on parade' – all of the same element. Diamond, gold, copper and iron are examples. But other crystals are made of two different kinds of atoms, again on parade in strict order: alternating, for example. Salt (common salt, table salt) is not an element but a compound of two elements, sodium and chlorine. In a crystal of salt, the sodium and chlorine atoms parade together alternately. Actually, in this case they are called not atoms but 'ions', but I'm not going to go into why that is. Every sodium ion has six chlorines for neighbours, at right angles to each other: in front, behind, to left, to right, above, and below. And every chlorine ion is surrounded by sodiums, in just the same way. The whole arrangement is composed of squares, and this is why salt crystals, if you look at them carefully with a strong lens, are cubic – the three-dimensional form of a square – or at least have squared-off edges. Lots of other crystals are made of more than one kind of atom 'on parade', and many of them are found in rocks, sand and soil.

Solid, liquid, gas – how molecules move

Crystals are solid, but not everything is solid. We also have liquids and gases. In a gas, the molecules don't stick together as they do in a crystal, but rush freely about within whatever space is available, travelling in straight lines like billiard balls (but in three dimensions, not two as on a flat table). They rush about until they hit something, such as another molecule or the walls of a con-

tainer, in which case they bounce off, again like billiard balls. Gases can be compressed, which shows there is a lot of space between the atoms and molecules. When you compress a gas, it feels 'springy'. Put your finger over the end of a bicycle pump and feel the springiness as you push the plunger in. If you keep your finger there, when you let the plunger go it shoots back out. The springiness that you are feeling is called 'pressure'. The pressure is the effect of all the millions of molecules of air (a mixture of nitrogen and oxygen and a few other gases) in the pump bombarding the plunger (and everything else, but the plunger is the only part that can move in response). At high pressure the bombardment happens at a higher rate. This will happen if the same number of gas molecules are confined in a smaller volume (for instance, when you push the plunger of a bicycle pump). Or it will happen if you raise the temperature, which makes the gas molecules charge about faster.

A liquid is like a gas in that its molecules move around or 'flow' (that's why both are called 'fluids', while solids aren't). But the molecules in a liquid are much closer to each other than the molecules in a gas. If you put a gas into a sealed tank, it fills every nook and cranny of the tank up to the top. The volume of gas rapidly expands to fill the whole tank. A liquid also fills every nook and cranny, but only up to a certain level. A given amount of liquid, unlike the same amount of gas, keeps a fixed volume, and gravity pulls it downwards, so it fills only as much as it needs of the tank, from the bottom upwards. That's because the molecules of a liquid stay close to each other. But, unlike those of a solid, they do slide around over each other, which is why a liquid behaves as a fluid.

A solid doesn't even try to fill the tank – it just retains its shape. That's because the molecules of a solid don't slide around over each other like those of a liquid, but stay in (roughly) the same positions relative to their neighbours. I say 'roughly' because even in a solid the molecules do sort of jiggle about (faster at higher temperatures): they just don't move far enough from their position in the crystal 'parade' to affect its shape.

Sometimes a liquid is 'viscous', like treacle. A viscous liquid flows, but so slowly that, although a very viscous liquid eventually fills the bottom part of the tank, it takes a long time to do so. Some liquids are so viscous – flow so slowly – that they might as well be solid. Substances of this kind behave like solids, even though they're not made of crystals. Glass is an example. Glass is said to 'flow', but so slowly that it takes centuries for us to notice. So, for practical purposes, we can treat glass as solid.

METHANE

8.72°C
MERCURY

535°C
IRON

Solid, liquid and gas are the names we give to the three common 'phases' of matter. Many substances are capable of being all three, at different temperatures. On Earth, methane is a gas (it's often called 'marsh gas', because it bubbles up from marshes, and sometimes it catches fire and we see it lit up as eerie 'will-o'-the-wisps'). But on a large, very cold moon of the planet Saturn called Titan there are lakes of liquid methane. If a planet were colder still, it might have 'rocks' of frozen methane. We think of mercury as a liquid, but that just means it's liquid at ordinary temperatures on Earth. Mercury is a solid metal if you leave it outside in the Arctic winter. Iron is a liquid if you heat it to a high enough temperature. Indeed, around the deep centre of the Earth is a sea of liquid iron mixed with liquid nickel. For all I know there may be very hot planets with oceans of liquid iron at the surface, and perhaps strange creatures swimming in them, although I doubt that. By our standards, the freezing point of iron is rather hot, so at the surface of the Earth we usually encounter it as 'iron – cold iron',[*] and the freezing point of mercury is rather cold, so we usually encounter it as 'quicksilver'. At the other end of the temperature scale, both mercury and iron become gases if you heat them enough.

Inside the atom

When we were imagining cutting matter into the smallest possible pieces at the beginning of this chapter, we stopped at the atom. An atom of lead is the smallest object that still deserves to be called lead. But can you really not cut an atom any further? And would an atom of lead actually look like a tiny little chip of lead? No, it wouldn't look like a tiny piece of lead. It wouldn't look like anything. That's because an atom is too small to be seen, even with a powerful microscope. And yes, you can cut an atom into even smaller pieces – but what you then get is no longer the same element, for reasons we shall soon see. What is more, this is very difficult to do, and it releases an alarming quantity of energy. That is why, for some people, the phrase 'splitting the atom' has such an ominous ring to it. It was first done by the great New Zealand scientist Ernest Rutherford in 1919.

[*] Google it. It's from the poet Rudyard Kipling, whom I love although he is rather unfashionable nowadays.

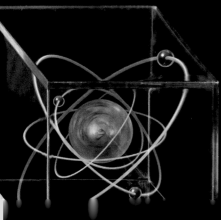

Although we can't see an atom, and although we can't split it without turning it into something else, that doesn't mean we can't work out what it is like inside. As I explained in Chapter 1, when scientists can't see something directly, they propose a 'model' of what it might be like, and then they test that model. A scientific model is a way of thinking about how things might be. So a model of the atom is a kind of mental picture of what the inside of an atom might be like. A scientific model can seem like a flight of fancy, but it is not just a flight of fancy. Scientists don't stop at proposing a model: they then go on to test it. They say, 'If this model that I am imagining were true, we would expect to see such-and-such in the real world.' They predict what you'll find if you do a particular experiment and make certain measurements. A successful model is one whose predictions come out right, especially if they survive the test of experiment. And if the predictions come out right, we hope it means that the model probably represents the truth, or at least a part of the truth.

Sometimes the predictions don't come out right, and so scientists go back and adjust the model, or think up a new one, and then go on to test that. Either way, this process of proposing a model and then testing it – what we call the 'scientific method' – has a much better chance of getting at the way things really are than even the most imaginative and beautiful myth invented to explain what people didn't – and often, at the time, couldn't – understand.

An early model of the atom was the so-called 'curran bun' model proposed by the great English physicist J. J. Thomson at the end of the nineteenth century. I won't describe it because it was replaced by the more successful Rutherford model, first proposed by the same Ernest Rutherford who split the atom, who came from New Zealand to England to work as Thomson's pupil and who succeeded Thomson as Cambridge's Professor of Physics. The Rutherford model, later refined in turn by Rutherford's pupil, the celebrated Danish physicist Niels Bohr, treats the atom as a tiny, miniaturized solar system. There is a nucleus in the middle of the atom, which contains the bulk of its material. And there are tiny particles called electrons whizzing around the nucleus in 'orbit' (though 'orbit' may be misleading if you think of it as just like a planet orbiting the sun, because an electron is not a little round thing in a definite place).

One surprising thing about the Rutherford / Bohr model, which probably reflects a real truth, is that the distance between each nucleus and the next is very large compared with the size of the nuclei, even in a hard chunk of solid matter like a diamond. The nuclei are hugely spaced out. This is the point I promised to return to.

Remember I said that a diamond crystal is a giant molecule made of carbon atoms like soldiers on parade, but a parade in three dimensions? Well, we can now improve our 'model' of the diamond crystal by giving it a scale – that is, a sense of how sizes and distances in it relate to one another. Suppose we represent the nucleus of each carbon atom in the crystal not by a soldier but by a football, with electrons in orbit around it. On this scale, the neighbouring footballs in the diamond would be more than 15 kilometres away.

The 15 kilometres between the footballs would contain the electrons in orbit around the nuclei. But each electron, on our 'football' scale, is much smaller than a gnat, and these miniature gnats are themselves several kilometres away from the footballs they are flying around. So you can see that – amazingly – even the legendarily hard diamond is almost entirely

e m p t y   s p a c e !

The same is true of all rocks, no matter how hard and solid. It is true of iron and lead. It is also true of even the hardest wood. And it is true of you and me. I've said that solid matter is made of atoms 'packed' together, but 'packed' means something rather odd here because the atoms themselves are mostly empty space. The nuclei of the atoms are spaced out so far apart that, if they were scaled up to footballs, any pair of them would be 15 kilometres apart with only a few gnats in between.

How can this be? If a rock is almost entirely empty space, with the actual matter dotted about like footballs separated by kilometres from their nearest neighbours, how come it feels so hard and solid? Why doesn't it collapse like a house of cards when you sit on it? Why can't we see right through it? If both a wall and I are mostly empty space, why can't I walk straight through the wall?

At first thought, it sounds plausible. I know that the wall, and my own body, are made of atoms so spaced out that they are like footballs 15 kilometres apart. Surely, if both the wall and

Imagine you're sitting in an ordinary room, in an ordinary building, gazing at the wall. You get to thinking: that wall's just concrete, and concrete is just atoms, and atoms are – mostly empty space. And so are you – mostly empty space.

Surely you could walk through the wall? Why not give it a go? So you do.

And then you hit the wall. Why?

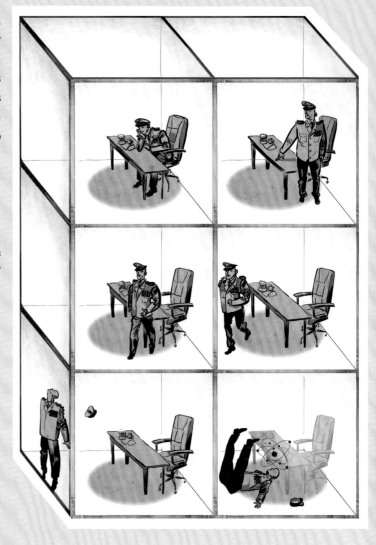

my own body are mostly empty space, I should be able to walk through the wall, slotting my atoms in between the wall's atoms? Why can't I?

Why do rocks and walls feel hard, and why can't we merge our spaces with theirs? We have to realize (as anyone who actually tries to walk through a wall will learn the hard way) that what we feel and see as solid matter is more than just nuclei and electrons – the 'footballs' and the 'gnats'. Scientists talk about 'forces' and 'bonds' and 'fields', which act in their different ways both to keep the 'footballs' apart and to keep the components of each 'football' together. And it is those forces and fields that make things feel solid.

When you get down to really small things like atoms and nuclei, the distinction between 'matter' and 'empty space' starts to lose its meaning. It isn't really right to say that the nucleus is 'matter' like a soccer ball,

and that there is 'empty space' until the next nucleus.

We define solid matter as 'what you can't walk through'. You can't walk through a wall because of these mysterious forces that link the nuclei to their neighbours in a fixed position. That's what solid means.

Liquid means something similar, except that the mysterious fields and forces hold the atoms together less tightly, so they slide over each other, which means that you can walk through water, although not so fast as you can walk through air. Air, being a gas (a mixture of gases, actually), is easy to walk through, because the atoms in a gas whizz about freely, rather than being tied to each other. A gas becomes difficult to walk through only if most of the atoms are whizzing in the same direction, and it is the opposite direction to the one in which you are trying to walk. This is what happens when you are trying to walk against the wind (that's what 'wind' means). It can be difficult to walk against a strong gale, and impossible against a hurricane or against the artificial gale hurled out behind a jet engine.

We can't walk through solid matter, but some very small particles such as the ones called photons can. Light beams are streams of photons, and they can go right through some kinds of solid matter – the kinds we call 'transparent'. Something about the way the 'footballs' are arranged in glass or in water or in certain gemstones means that

photons

can pass right between them, although they are slowed down a bit, just as you are slowed down when you try to walk through water.

With a few exceptions like quartz crystals, rocks aren't transparent, and photons can't pass through them. Instead, depending on the rock's colour, they are either absorbed by the rock or reflected from its surface, and the same is true of most other solid things. A few solid things reflect photons in a very special straight-line way, and we call them mirrors. But most solid things absorb many of the photons (they aren't transparent), and scatter even the ones that they reflect (they don't behave like mirrors). We just see them as 'opaque', and we also see them as having a colour, which depends on which kinds of photons they absorb and which kinds they reflect. I'll return to the important subject of colour in Chapter 7, 'What is a Rainbow?' Meanwhile, we need to shrink our vision to the very small indeed, and look right inside the nucleus – the football – itself.

The tiniest things of all

The nucleus isn't really like a football. That was just a crude model. It certainly isn't round like a football. It isn't even clear whether we should speak of it as having a 'shape' at all. Maybe the very word 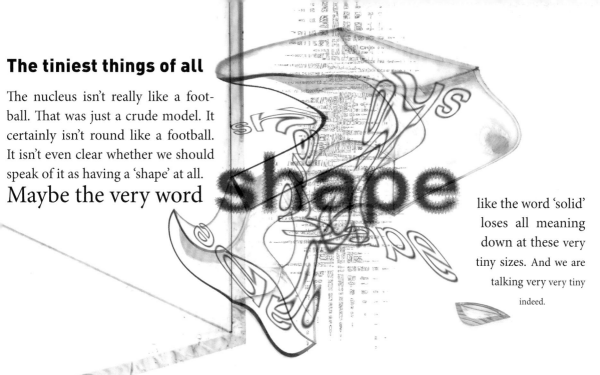 shape like the word 'solid' loses all meaning down at these very tiny sizes. And we are talking very very tiny indeed.

The full stop at the end of this sentence contains about a million million atoms of printing ink

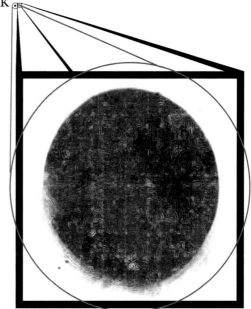

Each nucleus contains smaller particles called protons and neutrons. You can think of them as balls too, if you like, but like the nuclei they are not really balls. Protons and neutrons are approximately the same size as each other. They are very very tiny indeed, but even so they are still 1,000 times bigger than the electrons ('gnats') in orbit around the nucleus. The main difference between a proton and a neutron is that the proton has an electric charge. Electrons, too, have an electric charge, opposite to that of protons. We needn't bother with exactly what 'electric charge' means here. Neutrons have no charge.

Because electrons are so very very very tiny (while protons and neutrons are only very very tiny!) the mass of an atom is, to all intents and purposes, just its protons and neutrons. What does 'mass' mean? Well, you can think of mass as rather like weight, and you can measure it using the same units as weight (grams or pounds). Weight is not the same as mass, however, and I'll need to explain the difference, but I'm postponing that to the next chapter. For the moment just think of 'mass' as something like 'weight'.

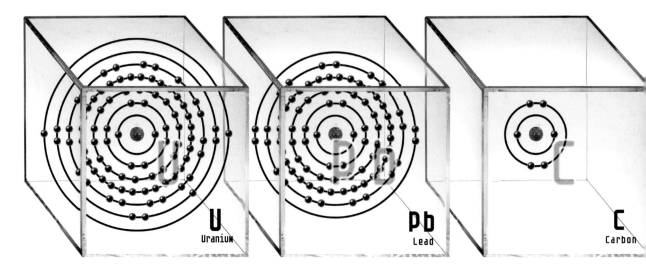

U
Uranium

Pb
Lead

C
Carbon

The mass of an object depends almost entirely on how many protons and neutrons it has in all its atoms added together. The number of protons in the nucleus of any atom of a particular element is always the same, and is equal to the number of electrons in orbit around the nucleus, although the electrons don't contribute noticeably to the mass because they are too small. A hydrogen atom has only one proton (and one electron). A uranium atom has 92 protons. Lead has 82. Carbon has 6. For every possible number from 1 to 100 (and a few more), there is one and only one element that has that number of protons (and

the same number of electrons). I won't list them all, but it would be easy to do so (Lalla, my wife, can recite them all by heart, at great speed, a trick she taught herself as an exercise in training her memory and as a device to help her get to sleep).

The number of protons (or electrons) that an element possesses is called the 'atomic number' of that element. So you can identify an element not just by its name but by its own unique atomic number. For example, element number 6 is carbon; element number 82 is lead. The elements are conveniently set out in a table called the periodic table – I won't go into why it's called that,

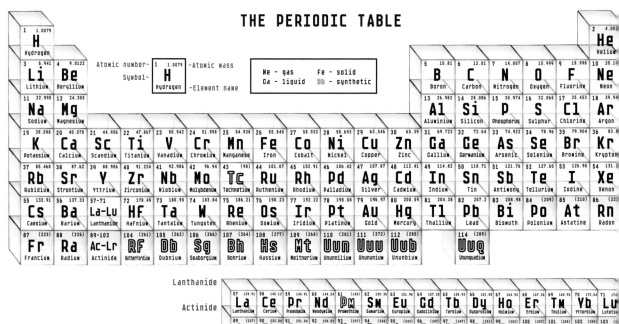

THE PERIODIC TABLE

although it is interesting. But now is the moment to return, as I promised I would, to the question of why, when you cut a piece of, say, lead into smaller and smaller pieces, you eventually reach a point where, if you cut it again, it is no longer lead. An atom of lead has 82 protons. If you split the atom so that it no longer has 82 protons it ceases to be lead.

The number of neutrons in an atom's nucleus is less fixed than the number of protons: many elements have different versions, called isotopes, with different numbers of neutrons. For example, there are three isotopes of carbon, called Carbon-12, Carbon-13 and Carbon-14. The numbers refer to the mass of the atom, which is the sum of the protons and neutrons. Each of the three has six protons. Carbon-12 has six neutrons, Carbon-13 has seven neutrons and Carbon-14 has eight neutrons. Some isotopes, for example Carbon-14, are radioactive, which means they change into other elements at a predictable rate, although at unpredictable moments. Scientists can use this feature to help them calculate the age of fossils. Carbon-14 is used to date things younger than most fossils, for example ancient wooden ships.

Well then, does our quest to cut things ever smaller and smaller end with these three particles: electrons, protons and neutrons? No – even protons and neutrons have an inside. Even they contain yet smaller things, called quarks. But that is something I'm not going to talk about in this book. That's not because I think you wouldn't understand it. It is because I know *I* don't understand it! We are here moving into a wonderland of the mysterious. And it is important to recognize when we reach the limits of what we understand. It is not that we shall never understand these things. Probably we shall, and scientists are working on them with every hope of success. But we have to know what we don't understand, and admit it to ourselves, before we can begin to work on it. There are scientists who understand at least something of this wonderland of the very small, but I am not one of them. I know my limitations.

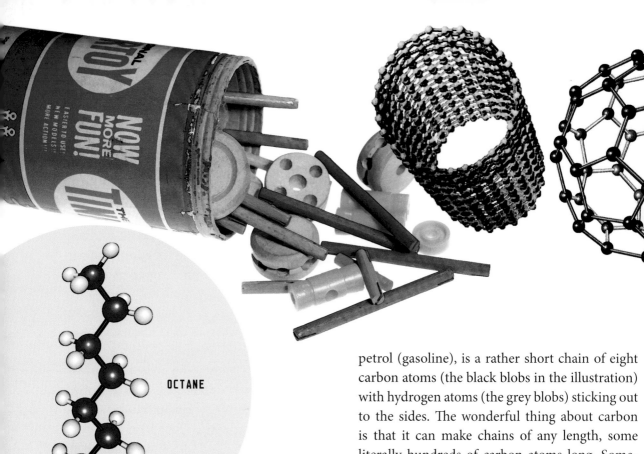

OCTANE

Carbon – the scaffolding of life

All the elements are special in their different ways. But one element, carbon, is so special that I want to end the chapter by talking briefly about that. Carbon chemistry even has its own name, separating it from the whole of the rest of chemistry: 'organic' chemistry. All the rest of chemistry is 'inorganic' chemistry. So what is so special about carbon?

The answer is that carbon atoms link up with other carbon atoms to form chains. The chemical compound octane (above), which, as you may know, is an ingredient of petrol (gasoline), is a rather short chain of eight carbon atoms (the black blobs in the illustration) with hydrogen atoms (the grey blobs) sticking out to the sides. The wonderful thing about carbon is that it can make chains of any length, some literally hundreds of carbon atoms long. Sometimes the chains come around in a loop. For example, above right is naphthalene (the substance that mothballs are made of), whose molecules are also made of carbon with hydrogen attached, this time in two loops. Carbon chemistry is rather like the toy construction kit called Tinkertoy.

In the laboratory, chemists have succeeded in making carbon atoms join up with each other, not just in simple loops but in wonderfully

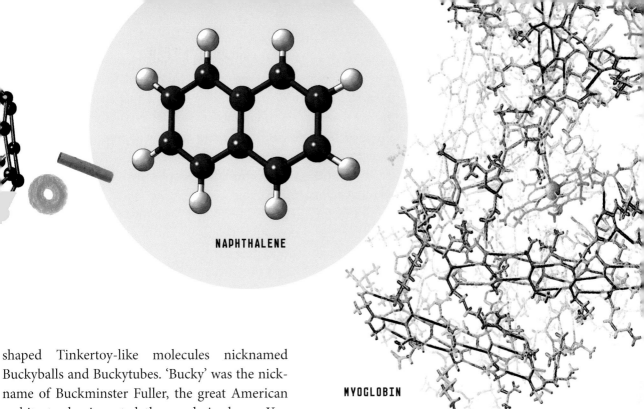

NAPHTHALENE

MYOGLOBIN

shaped Tinkertoy-like molecules nicknamed Buckyballs and Buckytubes. 'Bucky' was the nickname of Buckminster Fuller, the great American architect who invented the geodesic dome. You can see the connection if you look at the picture below. The Buckyballs and Buckytubes scientists have made are artificial molecules. But they show the Tinkertoyish way in which carbon atoms can be joined together into scaffolding-like structures that can be indefinitely large. (Just recently the exciting news was announced that Buckyballs have been detected in outer space, in the dust drifting near to a distant star.) Carbon chemistry offers a near-infinite number of possible molecules, all of different shapes, and thousands

of different ones are found in living bodies. Above is one very large molecule called myoglobin, which is found, in millions of copies, in all our muscles. The illustration does not show the individual atoms, just the bonds joining them.

Not all the atoms in myoglobin are carbon atoms, but it is the carbon atoms that join together in these fascinating Tinkertoy-like scaffolding structures. And that is really what makes life possible. When you think that myoglobin is only one example among thousands of equally complicated molecules in living cells, you can perhaps imagine that, just as you can build pretty much anything you like if you have a large enough Tinkertoy set, so the chemistry of carbon provides the vast range of possible forms required to put together anything so complicated as a living organism.

What, no myths?

This chapter has been unusual in that it didn't begin with a list of myths. This was only because it was so hard to find any myths on this subject. Unlike, say, the sun, or the rainbow, or earthquakes, the fascinating world of the very small never came to the notice of primitive peoples. If you think about this for a minute, it's not really surprising. They had no way of even knowing it was there, and so of course they didn't invent any myths to explain it! It wasn't until the microscope was invented in the sixteenth century that people discovered that ponds and lakes, soil and dust, even our own bodies, teem with tiny living creatures, too small to see, yet complicated and, in their own way, beautiful – or perhaps frightening, depending on how you think about them.

The creatures in the picture below are dust mites – distantly related to spiders but too small to see except as tiny specks. There are thousands of them in every home, crawling through every carpet and every bed, quite probably including yours.

If primitive peoples had known about them, you can imagine what myths and legends they might have invented to explain them! But before the invention of the microscope, their existence was not even dreamed of – and so there are no myths about them. And, small as it is, even a dust mite contains more than a hundred trillion atoms.

Dust mites are too small for us to see, but the cells of which they are made are smaller still. The bacteria that live inside them – and us – in vast numbers are smaller even than that.

And atoms are far far smaller even than bacteria. The whole world is made of incredibly tiny things, much too small to be visible to the naked eye – and yet none of the myths or so-called holy books that some people, even now, think were given to us by an all-knowing god, mentions them at all! In fact, when you look at those myths and stories, you can see that they don't contain any of the knowledge that science has patiently worked out. They don't tell us how big or how old the universe is; they don't tell us how to treat cancer; they don't explain gravity or the internal combustion engine; they don't tell us about germs, or nuclear fusion, or electricity, or anaesthetics. In fact, unsurprisingly, the stories in holy books don't contain any more information about the world than was known to the primitive peoples who first started telling them! If these 'holy books' really were written, or dictated, or inspired, by all-knowing gods, don't you think it's odd that those gods said nothing about any of these important and useful things?

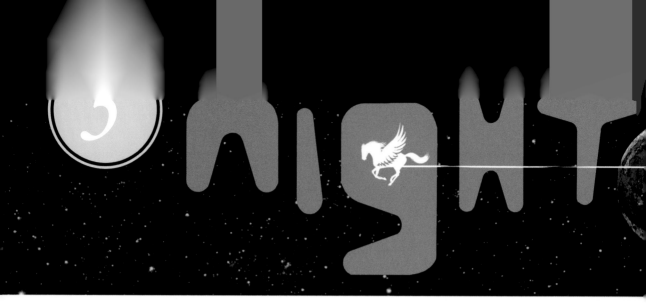

O UR LIVES are dominated by two great rhythms, one much slower than the other. The fast one is the daily alternation between dark and light, which repeats every 24 hours, and the slow one is the yearly alternation between winter and summer, which has a repeat time of a little over 365 days. Not surprisingly, both rhythms have spawned myths. The day–night cycle especially is rich in myth because of the dramatic way the sun seems to move from east to west. Several peoples even saw the sun as a golden chariot, driven by a god across the sky.

The aboriginal peoples of Australia were isolated on their island continent for at least 40,000 years, and they have some of the oldest myths in the world. These are mostly set in a mysterious age called the Dreamtime, when the world began and was peopled by animals and a race of giant ancestors. Different tribes of aborigines have different myths of the Dreamtime. This first one comes from a tribe who live in the Flinders Ranges of southern Australia.

During the Dreamtime, two liz-ards were friends. One was a goanna (the Australian name for a large monitor lizard) and the other a gecko (a delightful little lizard with suction pads on its feet, with which it climbs up vertical surfaces). The friends discovered that some other friends of theirs had been massacred by the 'sun-woman' and her pack of yellow dingo dogs.

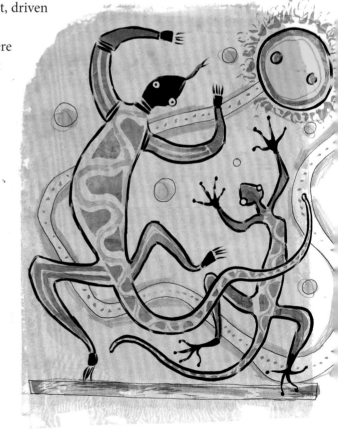

and day, winter and summer?

Furious with the sun-woman, the big goanna hurled his boomerang at her and knocked her out of the sky. The sun vanished over the western horizon and the world was plunged into darkness. The two lizards panicked and tried desperately to knock the sun back into the sky, to restore the light. The goanna took another boomerang and hurled it westwards, to where the sun had disappeared. As you may know, boomerangs are remarkable weapons that come back to the thrower, so the lizards hoped that the boomerang would hook the sun back up into the sky. It didn't. They then tried throwing boomerangs in all directions, in a vague hope of retrieving the sun. Finally, goanna lizard had only one boomerang left, and in desperation he threw it to the east, the opposite direction from where the sun had disappeared. This time, when it returned, it brought the sun with it. Ever since then, the sun has repeated the same pattern of disappearing in the west and reappearing in the east.

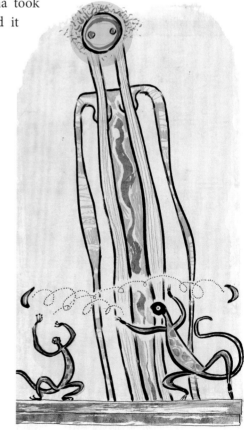

Many myths and legends from all around the world have the same odd feature: a particular incident happens once, and then, for reasons never explained, the same thing goes on happening again and again for ever.

Here's another aboriginal myth, this time from south-eastern Australia. Someone threw the egg of an emu (a sort of Australian ostrich) up into the sky. The sun hatched out of the egg and set fire to a pile of kindling wood which happened (for some reason) to be up there. The sky god noticed that the light was useful to men, and he told his servants to go out every night from then on, to put enough firewood in the sky to light up the next day.

The longer cycle of the seasons is also the subject of myths all around the world. Native North American myths, like many others, often have animal characters. In this one, from the Tahltan people of western Canada, there was a quarrel between Porcupine and Beaver over how long the seasons ought to be. Porcupine wanted winter to last five months, so he held up his five fingers. But Beaver wanted winter to last for more months than that – the number of grooves in his tail. Porcupine was angry and insisted on an even shorter winter. He dramatically bit off his thumb and held up the remaining four fingers. And ever since then winter has lasted four months. I find this a rather disappointing myth, because it already assumes that there will be a winter and summer, and explains only how many months each will last. The Greek myth of Persephone is better in this respect at least.

Persephone was the daughter of the chief god Zeus. Her mother was Demeter, fertility goddess of the Earth and the harvest. Persephone was greatly loved by Demeter, whom she helped in looking after the crops. But Hades, god of the underworld, home of the dead, loved Persephone too. One day, when she was playing in a flowery meadow, a great chasm opened up

and Hades appeared from below in his chariot; seizing Persephone, he carried her down and made her the queen of his dark, underground kingdom. Demeter was so grief-stricken at the loss of her beloved daughter that she stopped the plants growing, and people began to starve. Eventually Zeus sent Hermes, the gods' messenger, down to the underworld to fetch Persephone back up to the land of the living and the light. Unfortunately, it turned out that Persephone had eaten six pomegranate seeds while in the underworld, and this meant (by the kind of logic we have become used to where myths are concerned) that she had to go back to the underworld for six months (one for each pomegranate seed) in every year. So Persephone lives above ground for part of the year, beginning in the spring and continuing through summer. During this time, plants flourish and all is merry. But during the winter, when she has to return to Hades because she ate those pesky pomegranate seeds, the ground is cold and barren and nothing grows.

what **really** changes day to night, WINTER to SUMMER?

WHENEVER things change rhythmically with great precision, scientists suspect that either something is swinging like a pendulum or something is rotating: going round and round. In the case of our daily and seasonal rhythms, it's the second. The seasonal rhythm is explained by the yearly orbiting of the Earth around the sun, at a distance of about 93 million miles. And the daily rhythm is explained by the Earth's spinning round and round like a top.

The illusion that the sun moves across the sky is just that – an illusion. It's the illusion of *relative movement*. You will have met the same kind of illusion often enough. You are in a train, standing at a station next to another train. Suddenly you seem to start 'moving'. But then you realize that you aren't actually moving at all. It is the second train that is moving, in the opposite direction. I remember being intrigued by the illusion the first time I travelled in a train. (I must have been very young, because I also remember another thing I got wrong on that first train journey. While we were waiting on the platform, my parents kept saying things like 'Our train will be coming soon' and 'Here comes our train', and then 'This is our train now'. I was thrilled to get on it because this was *our* train. I walked up and down the corridor, marvelling at everything, and very proud because I thought we *owned* every bit of it.)

The illusion of relative movement works the other way, too. You think the other train has moved, only to discover that it is your own train that is moving. It can be hard to tell the difference between apparent movement and real movement. It's easy if your train starts with a jolt, of course, but not if your train moves very smoothly. When your train overtakes a slightly slower train, you can sometimes fool yourself into thinking your train is still and the other train is moving slowly backwards.

It's the same with the sun and the Earth. The sun is not really moving across our sky from east to west. What is really happening is that the Earth, like almost everything in the universe (including the sun itself, by the way, but we can ignore that), is spinning round and round. Technically we say the Earth is spinning on its 'axis': you can think of the axis as a bit like an axle running right through the globe from North Pole to South Pole. The sun stays almost still relative to the Earth (not relative to other things in the universe, but I am just going to write about how it seems to us here, on Earth). We spin too smoothly to feel the movement, and the air we breathe spins with us. If it didn't, we would feel it as a mighty rushing wind, because we spin at a thousand miles an hour. At least, that is the spin speed at the equator; obviously we spin more slowly as we approach the North or South Pole because the ground we're standing on has less far to go to complete a circuit round the axis. Since we can't feel the spinning of the planet, and the air spins with us, it's like the case of the two trains. The only way we can tell we are moving is to look at objects that are not spinning with us: objects like the stars and the sun. What we see is the relative movement, and – just as with the trains – it looks as though we are standing still and the stars and the sun are moving across our sky.

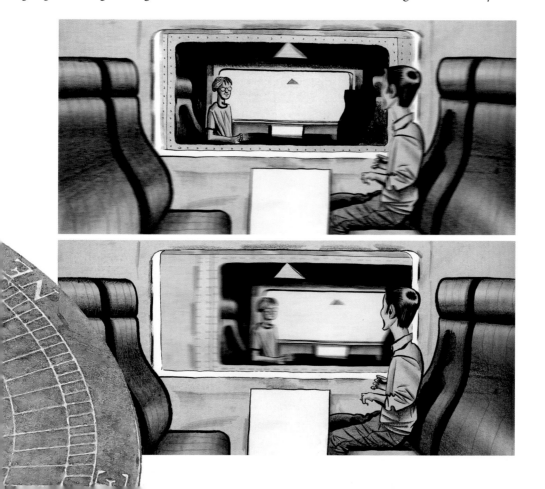

A famous thinker called Wittgenstein (the W is pronounced like a V) once asked a friend and pupil called Elizabeth Anscombe,

'Why do people say it was natural to think that the sun went round the Earth rather than that the Earth turned on its axis?'

Miss Anscombe answered,

'I suppose because it **looked** as if the sun went round the Earth.'

'Well,' Wittgenstein replied,

'what would it have looked like if it had looked as if the Earth turned on its axis?'

You try and answer that!

If the Earth is spinning at a thousand miles an hour, why, when we jump straight up in the air, don't we come down in a different place? Well, when you are on a train travelling at 100 mph, you can jump up in the air and you still land in the same place on the train. You can think of yourself as being hurled forwards by the train as you jump, but it doesn't feel like that because everything else is moving forwards at the same rate. You can throw a ball straight up on a train and it comes straight down again. You can play a perfectly good game of ping-pong on a train, so long as it is travelling smoothly and not accelerating or decelerating or going fast around a corner. (But only in an enclosed carriage. If you tried to play ping-pong on an open truck the ball would blow away. This is because the air comes with you in an enclosed carriage, but not when you are standing on an open truck.) When you are travelling at a steady rate in an enclosed railway carriage, no matter how fast, you might as well be standing stock still as far as ping-pong, or anything else that happens on the train, is concerned. However, if the train is speeding up (or slowing down), and you jump up in the air, you will come down in a different place! And a game of ping-pong on an accelerating or decelerating or turning train would be a strange game, even though the air inside the carriage is dead still relative to the carriage. We'll come back to this later, in connection with what it is like when you throw things about in an orbiting space station.

Working round the clock – and the calendar

Night gives way to day, and day gives way to night, as the part of the world we happen to be standing on spins to face the sun, or spins into the shade. But almost as dramatic, at least for those of us who live far from the equator, is the seasonal change from short nights and long, hot days in summer to long nights and short, cold days in winter.

The difference between night and day is dramatic – so dramatic that most species of animal can thrive either in the day or in the night but not both. They usually sleep during their 'off' period. Humans and most birds sleep by night and work at the business of living during the day. Hedgehogs and jaguars and many other mammals work by night and sleep by day.

In the same way, animals have different ways of coping with the change between winter and summer. Lots of mammals grow a thick, shaggy coat for the winter, then shed it in spring. Many birds, and mammals too, migrate, sometimes huge distances, to spend the winter closer to the equator, then migrate back to the high latitudes (the far north or far south) for the summer, where the long days and short nights provide bumper feeding. A seabird called the Arctic tern carries this to an extreme. Arctic terns spend the northern summer in the Arctic. Then, in the northern autumn, they migrate south – but they don't stop

in the tropics, they go all the way to the Antarctic. Books sometimes describe the Antarctic as the 'wintering grounds' of the Arctic tern, but of course that's nonsense: by the time they get to the Antarctic it is the southern summer. The Arctic tern migrates so far that it gets two summers: it has no 'wintering grounds' because it has no winter. I'm reminded of the joking remark of a friend of mine who lived in England during the summer, and went to tropical Africa to 'tough out the winter'!

Another way some animals avoid the winter is to sleep through it. It's called 'hibernation', from *hibernus,* the Latin word for 'wintry'. Bears and ground squirrels are among the many mammals, and quite a lot of other kinds of animals, that hibernate. Some animals sleep continuously through the whole winter; some sleep for most of the time, occasionally stirring into sluggish activity and then sleeping again. Usually their body temperature drops dramatically during hibernation and everything inside them slows down almost to a stop: their internal engines just barely tick over. There's even a frog in Alaska which goes so far as to freeze solid in a block of ice, thawing out and coming to life again in the spring.

Even those animals, like us, that don't hibernate or migrate to avoid the winter have to adapt to the changing seasons. Leaves sprout in spring and fall in autumn (which is why it's called the 'fall' in America), so trees that are a lush green in summer become gaunt and bare in winter. Lambs are born in spring, so they get the benefit of warm temperatures and new grass as they are growing. We may not grow long, woolly coats in winter, but we often wear them.

So we can't ignore the changing seasons, but do we understand them? Many people don't. There are even some people who don't understand that the Earth takes a year to orbit the sun – indeed, that's what a year *is*! According to a poll, 19 per cent of British people think it takes a month, and similar percentages have been found in other European countries.

Even among those who understand what a year means, there are many who think the Earth is closer

to the sun in summer, more distant in winter. Tell that to an Australian, barbecuing Christmas dinner in a bikini on a baking hot beach! The moment you remember that in the southern hemisphere December is midsummer and June is midwinter, you realize that the seasons can't be caused by changes in how close the Earth is to the sun. There has to be another explanation.

We can't get very far with that explanation until we have looked at what makes heavenly bodies orbit other heavenly bodies in the first place. So that's what we'll do next.

Into orbit

Why do the planets stay in orbit around the sun? Why does anything stay in orbit around anything else? This was first understood in the seventeenth century by Sir Isaac Newton, one of the greatest scientists who ever lived. Newton showed that all orbits were controlled by gravity – the same force of gravity that pulls falling apples towards the ground, but on a larger scale. (Alas, the story that Newton got the idea when an apple bounced off his head is probably not really true.)

Newton imagined a cannon on top of a very high mountain, with its barrel pointing horizontally out to sea (the mountain is on the coast). Each ball it fires seems to start off moving horizontally, but at the same time it is falling towards the sea. The combination of motion out over the sea and falling towards the sea results in a graceful downward curve, culminating in a splash. It is important to understand that the ball is falling all the time, even on the earlier, flatter part of the curve. It's not that it travels flat horizontally for a while, then suddenly changes its mind like a cartoon character who realizes he ought to be falling and therefore starts doing so!

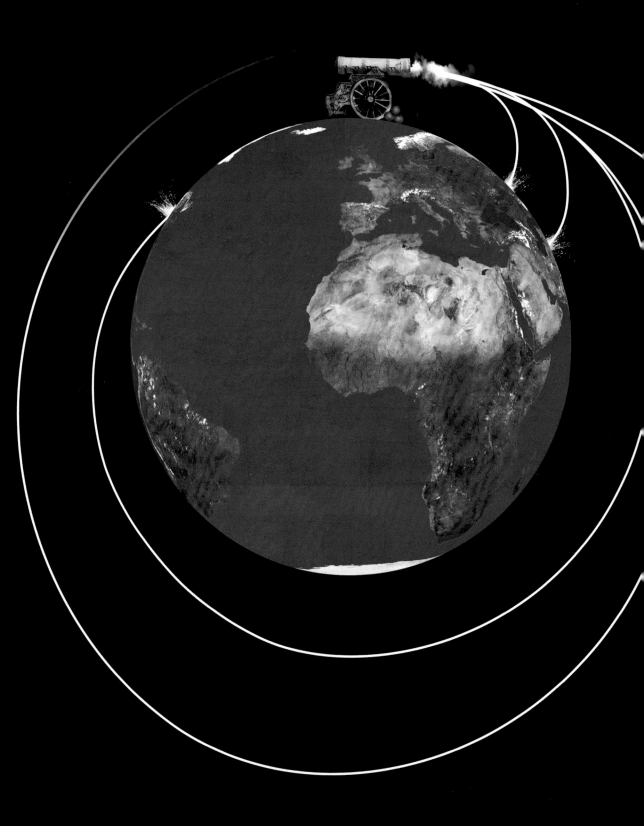

The cannonball starts falling the moment it leaves the gun, but you don't see the falling as downward motion because the ball is moving (nearly) horizontally as well, and quite fast.

Now let's make our cannon bigger and stronger, so that the cannonball travels many miles before it finally splashes into the sea. There is still a downward curve, but it's a very gradual, very 'flat' curve. The direction of travel is pretty nearly horizontal for quite a lot of the way, but nevertheless it is still falling the whole time.

Let's carry on imagining a bigger and bigger cannon, more and more powerful: so powerful that the ball travels a really long way before it goes into the sea. Now the curvature of the Earth starts to make itself felt. The ball is still 'falling' the whole time, but because the planet's surface is curved, 'horizontal' now starts to mean something a bit odd. The cannonball still follows a graceful curve, as before. But as it slowly curves towards the sea, the sea curves away from it because the planet is round. So it takes even longer for the cannonball finally to splash down into the sea. It is still falling all the time, but it is falling *around* the planet.

You can see the way the argument is going. We now imagine a cannon so powerful that the ball keeps going all the way around the Earth till it arrives back where it started. It is still 'falling', but the curve of its fall is matched by the curvature of the Earth so that it goes right round the planet without getting any closer to the sea. It is now *in orbit* and it will keep on orbiting the Earth for an indefinite time, assuming that there is no air resistance to slow it down (which in reality there would be). It will still be 'falling', but the graceful curve of its prolonged fall will go all around the Earth, and around again and again. It will behave just like a miniature moon. In fact, that

is what satellites are – artificial moons. They are all 'falling' but they never actually come down. The ones that are used for relaying long-distance telephone calls or television signals are in a special orbit called a geostationary orbit. This means that the rate at which they go around the Earth has been cunningly arranged so that it is exactly the same as the rate at which the Earth spins on its own axis: that is, they orbit the Earth once every 24 hours. This means, if you think about it, that they are always hovering above exactly the same spot on the Earth's surface. That is why you can aim your satellite dish precisely at the particular satellite that is beaming down the television signal.

When an object, such as a space station, is in orbit, it is 'falling' the whole time, and all the objects in the space station, whether we think of them as light or heavy, are falling at the same rate. This is a good place to stop a moment and explain the difference between mass and weight, as I promised to do back in the previous chapter.

All objects in an orbiting space station are weightless. But they are not massless. Their mass, as we saw in that chapter, depends on the number of protons and neutrons they contain. Weight is the pull of gravity on your mass. On Earth we can use weight to measure mass because the pull is (more or less) the same everywhere. But because more massive planets have stronger gravity, your weight changes depending which planet you are on, whereas your mass stays the same wherever you are – even if you are completely weightless in a space station in orbit. You'd be weightless on the space station because you and the weighing machine would both be 'falling' at the same rate (in what is called 'free fall'); so your feet would exert no pressure on the weighing machine, which would therefore register you as weightless.

But although you'd be weightless, you'd be far from massless. If you jumped vigorously away from the 'floor' of the space station, you'd shoot towards the 'ceiling' (it wouldn't be obvious which was floor and which ceiling!) and, no matter how far away the ceiling was, you'd bang your head and it would hurt, just as if you had fallen on your head. And everything else in the space station would still have its own mass likewise. If you had a cannonball in the cabin with you, it would float about weightlessly, which might make you think it was light like a beach ball of the same size. But if you tried to throw it across the cabin, you'd soon know that it wasn't light like a beach ball. It would be hard work to throw it, and you might find yourself shooting backwards in the opposite direction if you tried. The cannonball would feel heavy, even though it would show no special tendency to go 'downwards' towards the floor of the space station. If you succeeded in throwing the cannonball across the room, it would behave like any heavy object when it hit something in its path, and it would not be good if it hit one of your fellow astronauts on the head, either directly or after bouncing off the wall. If it hit another cannonball, the two would bounce off each other with a proper 'heavy' feel, unlike, say, a pair of ping-pong balls, which would also bounce off each other but lightly. I hope that gives you a feel for the difference between weight and mass. In the space station, a cannonball has much more mass than a balloon, although both have the same weight – zero.

Eggs, ellipses and escaping gravity

Let's go back to our cannon on the mountain-top, and make it more powerful still. What will happen? Well, now we need to acquaint ourselves with the discovery of the great German scientist Johannes Kepler, who lived just before Newton. Kepler showed that the graceful curve by which things orbit other things in space is not really a circle but something known to mathematicians since ancient Greek times as an 'ellipse'. An ellipse is sort of egg-shaped (only 'sort of': eggs are not perfect ellipses). A circle is a special case of an ellipse – think of a very blunt egg, an egg so short and squat that it looks like a ping-pong ball.

There's an easy way to draw an ellipse, while at the same time convincing yourself that a circle is a special case of an ellipse. Take a piece of string and make it into a loop by tying the ends together, in as neat and small a knot as you can. Now stick a pin in a pad of paper, loop the string around the pin, stick a pencil through the other end of the loop, pull it tight and draw all around the pin with the string loop at full stretch. You'll draw a circle, of course.

Next, take a second pin and stick it in the pad, right next to the first pin so that they are touching. You'll still draw a circle because the two pins are so close together that they count as a single pin. But now here's the interesting part. Move the pins apart a few inches. Now when you draw with the string at full stretch, the shape you produce will not be a circle, it will be an 'egg-shaped' ellipse. The further apart you place the two pins, the narrower the ellipse will be. The closer you place the two pins to each other, the wider – the more circular – the ellipse will be until, when the two pins become one pin, the ellipse will be a circle – the special case.

Now that we have met the ellipse we can go back to our super-powerful cannon. It has already fired a cannonball into an orbit which we assumed to be nearly circular. If we now make it more powerful still, what happens is that the orbit becomes a more 'stretched', less circular ellipse. This is called an 'eccentric' orbit. Our cannonball zooms quite a long way from the Earth, then turns around and falls back. Earth is one of the two 'pins'. The other 'pin' doesn't really exist as a solid object, but you can think of it as an imaginary pin out there in space. The imaginary pin helps to make the mathematics understandable for some people but if it confuses you just forget about it. The important thing to realize is that the Earth is not in the centre of the 'egg'. The orbit stretches much further away from the Earth on one side (the side of the 'imaginary pin') than on the other (the side where the Earth itself is the 'pin').

We go on making our cannon more and more powerful. The cannonball is now travelling a long, long way from the Earth and is only just pulled back around to fall back towards Earth. The ellipse is now very long and stretched indeed. And there will eventually come a point where it ceases to be an ellipse altogether: we fire the cannonball even faster, and now the extra speed just pushes it beyond the point of no return, where the Earth's gravity can't summon it back. It has reached 'escape velocity' and disappears for ever (or until captured by the gravity of another body, such as the sun).

Our increasingly powerful cannon has illustrated all the stages towards and beyond the establishment of an orbit. First the ball just flops into the sea. Then, as we fire successive balls with increasing force, the curve of their travel becomes increasingly horizontal until the ball reaches the necessary speed to go into a near-circular orbit (remember that a circle is a special case of an ellipse). Then, as the speed of firing increases more and more, the orbit becomes less circular and more elongated, more obviously elliptical. Finally, the 'ellipse' becomes so elongated that it ceases to be an ellipse at all: the ball reaches escape velocity and disappears altogether.

The Earth's orbit around the sun is technically an ellipse, but it is very nearly the special case of a circle. The same is true of all the other planets except Pluto (which is not considered a planet nowadays anyway). A comet, on the other hand, has an orbit like a very long, thin egg. The 'pins' that you use to draw its ellipse are very far apart.

One of the two 'pins' for a comet is the sun. Once again, the other 'pin' is not a real object in space: you just have to imagine it. When a comet is at its furthest distance from the sun (called 'aphelion', pronounced app-heeleeon) it travels at its slowest rate. It is in free fall the whole time, but some of the time it is falling away from the sun, rather than towards it. Slowly it turns the corner at aphelion, then it falls in the direction of the sun, falling faster and faster until it zooms round the sun (the other 'pin') and reaches its highest speed when it is at its closest point to the sun, called perihelion. ('Perihelion' and 'aphelion' come from the name of the Greek sun god Helios; *peri* is the Greek for 'near' and *apo* means 'far'.) The comet whizzes fast around the sun at perihelion, and carries on away from it at high speed on the other side of perihelion. After slinging itself around the sun, the comet gradually loses speed as it falls away from the sun all the way to aphelion, where it is at its slowest; and the cycle keeps repeating itself over and over again.

Space engineers use something called the slingshot effect to improve the fuel economy of their rockets. The Cassini space probe, which was designed to visit the distant planet Saturn, travelled there by what seems like a roundabout route, but was actually cunningly planned to exploit the slingshot effect. Using far less rocket fuel than would have been needed to fly directly to Saturn, Cassini borrowed from the gravity and orbital movement of three planets on the way – Venus (twice), then a return swing around Earth, then a final mighty heave from Jupiter. In each case it fell around the planet like a comet, gaining speed by hanging onto its gravitational coat-tails as the planet whizzed around the sun. These four slingshot boosts hurled Cassini out towards the Saturn system of rings and 62 moons, from

where it has been sending back stunning pictures ever since.

Most of the planets, as I said, orbit the sun in near-circular ellipses. Pluto is unusual, not just in being too small to be called a planet any more, but also in having a noticeably eccentric orbit. Much of the time it is outside the orbit of Neptune, but at perihelion it swoops inside and is actually closer to the sun than is Neptune, with its near-circular orbit. Even the orbit of Pluto, however, is nothing like as eccentric as that of a comet. The most famous one, Halley's Comet, becomes visible to us only near perihelion, when it is closest to the sun and reflects the sun's light. Its elliptical orbit takes it far, far away, and it returns to our neighbourhood only every 75 to 76 years. I saw it in 1986 and showed it to my baby

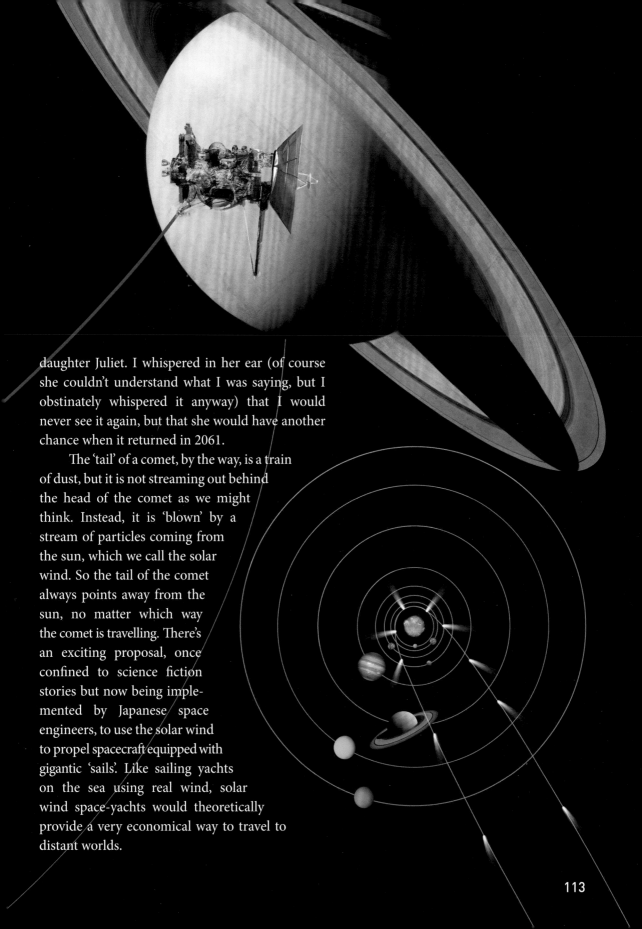

daughter Juliet. I whispered in her ear (of course she couldn't understand what I was saying, but I obstinately whispered it anyway) that I would never see it again, but that she would have another chance when it returned in 2061.

The 'tail' of a comet, by the way, is a train of dust, but it is not streaming out behind the head of the comet as we might think. Instead, it is 'blown' by a stream of particles coming from the sun, which we call the solar wind. So the tail of the comet always points away from the sun, no matter which way the comet is travelling. There's an exciting proposal, once confined to science fiction stories but now being implemented by Japanese space engineers, to use the solar wind to propel spacecraft equipped with gigantic 'sails'. Like sailing yachts on the sea using real wind, solar wind space-yachts would theoretically provide a very economical way to travel to distant worlds.

A sideways look at summer

Now that we understand orbits, we can go back to the question of why we have winter and summer. Some people, you'll remember, wrongly think it is because we are closer to the sun in summer and further away in winter. That would be a good explanation if Earth had an orbit like Pluto's. In fact Pluto's winter and summer (both very much colder than anything we experience here) are caused in exactly that way.

The Earth's orbit, however, is almost circular, so the planet's closeness to the sun cannot be what causes the changing seasons. For what it is worth, the Earth is actually closest to the sun (perihelion) in January and furthest (aphelion) in July, but the elliptical orbit is so close to circular that it makes no noticeable difference.

Well then, what does cause the difference between winter and summer? Something quite different. The Earth spins on an axis, and the axis is tilted. This tilting is the true reason why we have seasons. Let's see how it works.

As I said before, we could think of the axis as an axle, a rod running right through the globe and sticking out at the North Pole and the South Pole. Now think of the orbit of the Earth around the sun as a much larger wheel, with its own axle, this time running through the sun, and sticking out at the sun's 'north pole' and the sun's 'south pole'. Those two axles could have been exactly parallel to each other, so that the Earth did not have a 'tilt' – in which case the noonday sun would always seem to be directly overhead at the equator, and day and night would be of equal length everywhere. There would be no seasons. The equator would be perpetually hot, and it would become colder and colder the further you moved away from the equator and towards either of the poles. You could get cool by moving away from the equator, but not by waiting for winter because there would be no winter to wait for. No summer, no seasons of any kind.

In fact, however, the two axles are not parallel. The axle (axis) of the Earth's own spinning is tilted relative to the axle (axis) of our orbit around the sun. The tilt is not particularly great – about 23.5 degrees. If it were 90 degrees (which is about the tilt of the planet Uranus) the North Pole would be pointing straight towards the sun at one time of year (which we can call the northern midsummer) and straight away from the sun at the northern midwinter. If Earth were like Uranus, in midsummer the sun would be overhead all the time at the North Pole (there'd be no night there), while it would be icy cold and dark at the South Pole, with no suggestion of day. And vice versa six months later.

Since our planet is actually tilted at only 23.5 degrees instead of 90 degrees, we are about a quarter of the way from the no-seasons extreme of no tilt at all towards the Uranus extreme of near-total tilt. This is enough to mean that, as on Uranus, the sun never sets at the Earth's North Pole in midsummer. It is perpetual day; but, unlike on Uranus, the sun is not overhead. It seems to loop around the sky as the Earth

rotates, but it never quite dips below the horizon. That is true throughout the Arctic Circle. If you stood right on the Arctic Circle, say on the north-west tip of Iceland, on midsummer day, you'd see the sun skim along the northern horizon at midnight, but never actually set. Then it would loop around to its highest position (not very high) at midday.

In northern Scotland, which is a little way outside the Arctic Circle, the midsummer sun dips below the horizon far enough to make a sort of night – but not a very dark night, because the sun is never very far below the horizon.

So, the tilt of the Earth's axis explains why we have winter (when the bit of the planet where we are is tilted away from the sun) and summer (when it's tilted towards the sun), and why we have short days in winter and long days in summer. But does that explain why it is so cold in winter and so hot in summer? Why does the sun feel hotter when it is directly overhead than when it is low, near the horizon? It's the same sun,

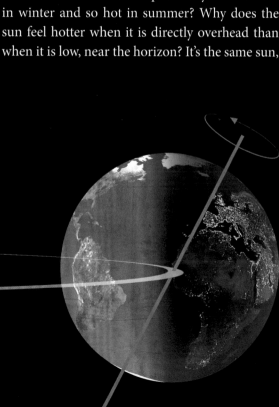

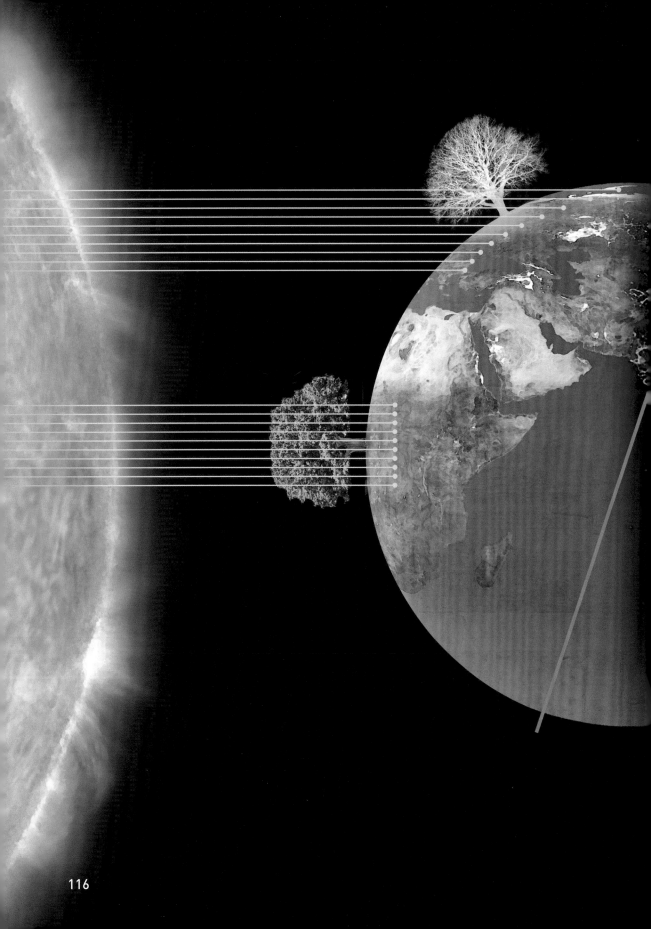

so shouldn't it be equally hot no matter what the angle at which we see it? No.

You can forget the fact that we are slightly nearer the sun when tilted towards it. That's an infinitesimal difference (only a few thousand miles) compared to the total distance from the sun (about 93 million miles), and still negligible compared to the difference between the sun's distance at perihelion and the sun's distance at aphelion (about 3 million miles). No, what matters is partly the angle at which the sun's rays hit us, and partly the fact that the days are longer in summer and shorter in winter. It's that *angle* that makes the sun feel hotter at midday than in the late afternoon, and it's that angle that makes it more important to put on sunscreen at midday than in the late afternoon. It's a combination of the angle and the day length that makes the plants grow more in summer than in winter, with all that follows from that.

So why does this angle make such a difference? Here's one way to explain it. Imagine that you are sunbathing at midday in the middle of the summer, and the sun is high overhead. A particular square inch of skin in the middle of your back is being hit by photons (tiny particles of light) at a rate that you could count with a light meter. Now, if you sunbathe at midday in winter, when the sun is relatively low in the sky because of the Earth's tilt, light reaches the Earth at a shallower, more 'sideways' angle: therefore a given number of photons are 'shared out' over a larger area of skin. This means that the original square inch of skin gets a smaller share of the available photons than it did at midsummer. What is true of your skin is also true of the leaves of plants, and that really matters because plants use sunlight to make their food.

Night and day, winter and summer: these are the great alternating rhythms that rule our lives, and the lives of all living creatures – except perhaps those that live in the dark, cold depths of the sea. Another set of rhythms that are not so important for us but matter greatly to other creatures, such as those that live on seashores, are the rhythms imposed by the orbiting moon, acting mostly through the tides. Lunar cycles are also the subject of ancient and disturbing myths – of werewolves and vampires, for example. But I must reluctantly leave this subject now and move on to the sun itself.

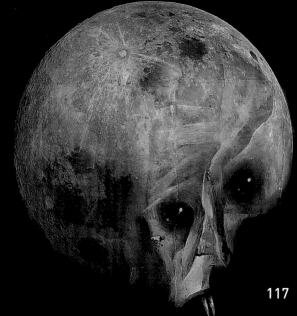

6 WHAT IS THE SUN?

THE SUN is so dazzlingly bright, so comforting in cold climates, so mercilessly scorching in hot ones, it is no wonder many peoples have worshipped it as a god. Sun worship often goes together with moon worship, and the sun and the moon are frequently regarded as being of opposite sex. The Tiv tribe of Nigeria and other parts of west Africa believe the sun is the son of their high god Awondo, and the moon is Awondo's daughter. The Barotse tribe of south-east Africa think the sun is the moon's husband rather than her brother. Myths often treat the sun as male and the moon female, but it can be the other way around. In the Japanese Shinto religion the sun is the goddess Amaterasu, and the moon is her brother Ogetsuno.

Those great civilizations that flourished in South and Central America before the Spaniards arrived in the sixteenth century worshipped the sun. The Inca of the Andes believed that the sun and the moon were their ancestors. The Aztecs of Mexico shared many of their gods with older civilizations in the area, such as the Maya. Several of these gods had a connection with the sun, or in some cases were the sun. The Aztec 'Myth of the Five Suns' held that there had been four worlds before the present one, each with its own sun. The earlier four worlds were destroyed, one after the other, by catastrophes, often engineered by the gods. The first sun was the god called Black Tezcatlipoca; he fought with his brother, Quetzalcoatl, who knocked him out of the sky with his club. After a period of darkness, with no sun, Quetzalcoatl became the second sun. In his anger, Tezcatlipoca turned all the people into monkeys, where-upon Quetzalcoatl blew all the monkeys away, and then resigned as the second sun.

The god Tlaloc then be-came the third sun. Annoyed when Tezcatlipoca stole his wife Xochiquetzal, he sulked and refused to allow any rain to fall, so there was a terrible drought. The people begged and begged for rain, and Tlaloc became so fed up with their begging that he sent down a rain of fire instead. This burned up the world, and the gods had to start all over again.

119

The fourth sun was Tlaloc's new wife, Chalchiuhtlicue. She started out well, but then Tezcatlipoca so upset her that she cried tears of blood for 52 years without stopping. This completely flooded the world, and yet again the gods had to start from scratch. Isn't it strange, by the way, how exactly myths specify little details? How did the Aztecs decide that she cried for 52 years, not 51 or 53?

The fifth sun, which the Aztec believed is the present one tha we still see in the sky, was th god Tonatiuh, sometimes know as Huitzilopochtli. His mothe Coatlicue, gave birth to hir after being accidentally impreg nated by a bundle of feathers This might sound odd, but suc things would have seemed quit normal to people brought up wit traditional myths (another Azte goddess was impregnated by gourd, which is the dried ski of a fruit like a pumpkin). Coat licue's 400 sons were so enrage to find their mother pregnant ye again that they tried to behea her. However, in the nick of tim she gave birth to Huitzilopochtl He was born fully armed and los no time in killing all of his 40 half-brothers, except a few wh escaped 'to the south'. Huitzilo pochtli then assumed his dutie as the fifth sun.

The Aztecs believed that they had to sacrifice human victims to appease the sun god, otherwise he would not rise in the east each morning. Apparently it didn't occur to them to try the experiment of not making sacrifices, to see whether the sun might, just possibly, rise anyway. The sacrifices themselves were famously gruesome. By the end of the Aztecs' heyday, when the Spaniards arrived (bringing their own brand of gruesomeness), the sun cult had escalated to a gory climax. It is estimated that between 20,000 and 80,000 humans were sacrificed for the rededication of the Great Temple of Tenochititlan in 1487. Various gifts could be offered to appease the sun god, but what he really liked was human blood, and still-beating human hearts. One of the main purposes of warfare was to collect lots of prisoners of war so that they could be sacrificed, usually by having their hearts cut out. The ceremony normally took place on high ground (to be closer to the sun), for example on top of one of the magnificent pyramids for which the Aztecs, Maya and Inca are famous. Four priests would hold the victim down over the altar, while a fifth priest wielded the knife. He worked as fast as possible to cut the heart out so that it was still beating when held up to the sun. Meanwhile the heartless and bloody corpse would roll down the slopes of the hill or pyramid to the bottom, where it would be collected up by the old men and then dismembered, often to be eaten in ritual meals.

We also associate pyramids with another ancient civilization, that of Egypt. The ancient Egyptians, too, were sun-worshippers. One of the greatest of their gods was the sun god Ra.

An Egyptian legend regarded the curve of the sky as the body of the goddess Nut, arched over the Earth. Every night the goddess swallowed the sun, and then the following morning she gave birth to him again.

Various peoples, including the ancient Greeks and the Norsemen, had legends about the sun being a chariot driven across the sky. The Greek sun god was called Helios, and he has given his name to various scientific terms associated with the sun, as we saw in Chapter 5.

In other myths, the sun is not a god but one of the first creations of a god. In the creation myth of the Hebrew tribe of the Middle Eastern desert, the tribal god YHWH created light on the first of his six days of creation – but then, surprisingly, he didn't create the sun until the fourth day! 'And God made two great lights: the greater light to rule the day, and the lesser light to rule the night: he made the stars also.' Where the light came from on the first day, before the sun and stars existed, we are not told.

It is time to turn to reality, and the true nature of the sun, as borne out by scientific evidence.

What is the sun, really?

The sun is a star. It's no different from lots of other stars, except that we happen to be near it so it looks much bigger and brighter than the others. For the same reason, the sun, unlike any other star, feels hot, damages our eyes if we look straight at it, and burns our skin red if we stay out in it too long. It is not just a *little* bit nearer than any other star; it is vastly nearer. It is hard to grasp how far away the stars are, how big space is. Actually, it's more than hard, it's almost impossible.

There's a lovely book called *Earthsearch* by John Cassidy, which makes an attempt to grasp it, using a scale model.

1 Go out into a big field with a football and plonk it down to represent the sun.

2 Then walk 25 metres away and drop a peppercorn to represent the Earth's size and its distance from the sun.

3 The moon, to the same scale, would be a pinhead, and it would be only 5 centimetres away from the peppercorn.

4 But the nearest other star, Proxima Centauri, to the same scale, would be another (slightly smaller) football located about . . . wait for it . . .

. . . 6,500 kilometres away!

There may or may not be planets orbiting Proxima Centauri, but there certainly are planets orbiting other stars, maybe most stars. And the distance between each star and its planets is usually small compared to the distance between the stars themselves.

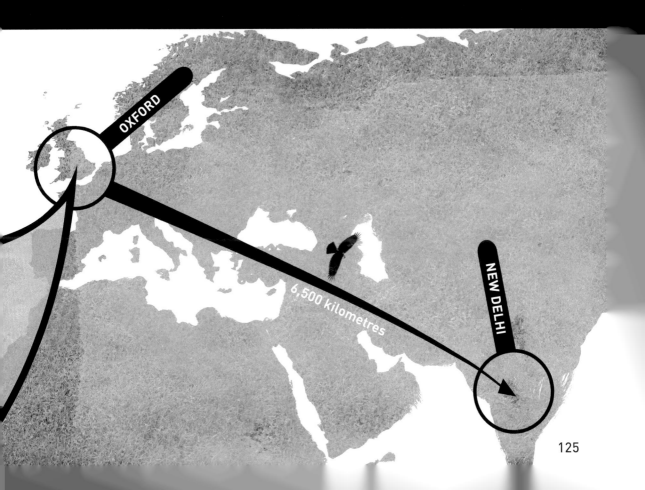

OXFORD

6,500 kilometres

NEW DELHI

How stars work

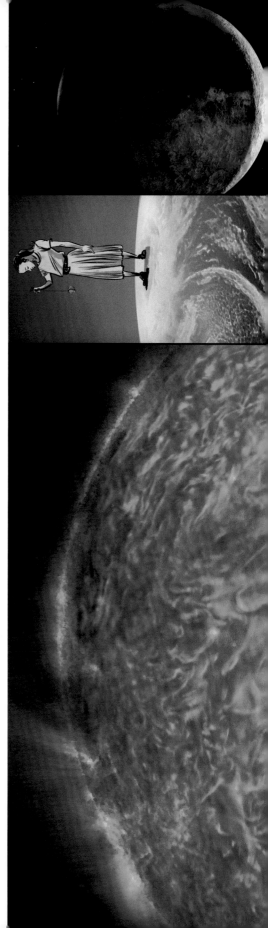

The difference between a star (like the sun) and a planet (like Mars or Jupiter) is that stars are bright and hot, and we see them by their own light, whereas planets are relatively cold and we see them only by reflected light from a nearby star, which they are orbiting. And that difference, in turn, results from the difference in size. Here's how.

The larger any object is, the stronger the gravitational pull towards its centre. Everything pulls everything by gravity. Even you and I exert a gravitational pull on each other. But the pull is too weak to notice unless at least one of the bodies concerned is large. The Earth is large, so we feel a strong pull towards it, and when we drop something it falls 'downwards' – that is, towards the centre of the Earth.

A star is much larger than a planet like Earth, so its gravitational pull is much stronger. The middle of a large star is under huge pressure because a gigantic gravitational force is pulling all the stuff in the star towards the centre. And the greater the pressure inside a star, the hotter it gets. When the temperature gets really high – much hotter than you or I can possibly imagine – the star starts to behave like a sort of slow-acting hydrogen bomb, giving out huge quantities of heat and light, and we see it shining brightly in the night sky. The intense heat tends to make the star swell up like a balloon, but at the same time gravity pulls it back in again. There is a balance between the outward push of the heat and the inward pull of gravity. The star acts as its own thermostat. The hotter it gets, the more it swells; and the bigger it gets, the less concentrated the mass of matter in the centre becomes, so it cools down a bit. This means it starts to shrink again, and that heats it up again, and so on. I've told the story as though the star bounces in and out like a heart beating, but it isn't like that. Instead, it settles into an intermediate size, which keeps the star at just the right temperature to stay that way.

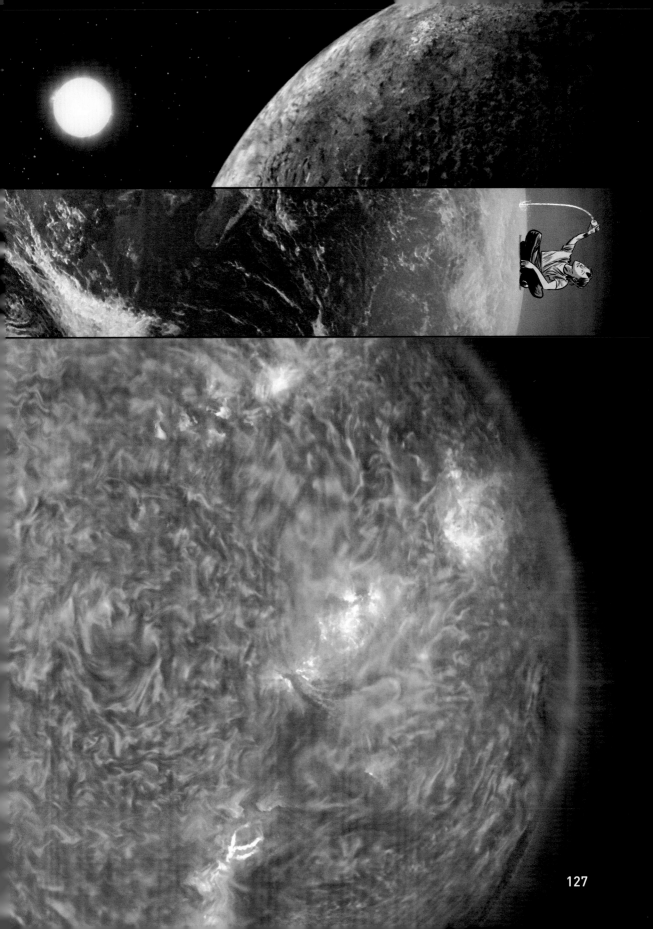

I began by saying that the sun is just a star like many others, but actually there are lots of different kinds of stars, and they come in a great range of sizes. Our sun (below) is not very big, as stars go. It is slightly bigger than Proxima Centauri, but much smaller than lots of other stars.

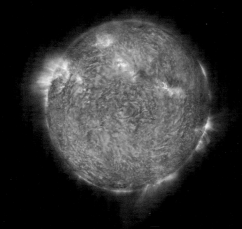

What is the largest star we know? That depends on how you measure them. The star that measures the greatest distance across is called VY Canis Majoris. From side to side (diameter), it is 2,000 times the size of the sun. And the sun's diameter is 100 times that of the Earth. However, VY Canis Majoris is so wispy and light that, despite its huge size, its mass is only about 30 times that of the sun, instead of the billions of times it would be if its material were equally dense. Others, such as the Pistol Star, and more recently discovered stars such as Eta Carinae and R136a1 (not a very catchy name!), are 100 times as massive as the sun, or even more. And the sun is more than 300,000 times the mass of the Earth, which means that the mass of Eta Carinae is 30 million times that of the Earth.

If a giant star like R136a1 has planets, they must be very very far away from it, or they would be instantly burned to vapour. Its gravity is so huge (because of its vast mass) that its planets could indeed be a very long way away and still be held in orbit around it. If there is such a planet, and anybody lives on it, R136a1 would probably look about as big to them as our sun looks to us, because although it is much larger, it would also be much further away – just the right distance away, in fact, and just the right apparent size to sustain life, otherwise life wouldn't be there!

The life story of a star

Actually, however, it is unlikely that there are any planets orbiting R136a1, let alone any life on them. The reason is that extremely large stars have a very short life. R136a1 is probably only about a million years old, which is less than a thousandth of the age of the sun so far: not enough time for life to evolve.

The sun is a smaller, more 'mainstream' star: the kind of star that has a life story lasting billions of years (not just millions), during which it proceeds through a series of drawn-out stages, rather like a child growing up, becoming an adult, passing through middle age, eventually getting old and dying. Mainstream stars mostly consist of hydrogen, the simplest of all the elements (see Chapter 4). The 'slow-acting hydrogen bomb'

in the interior of a star converts hydrogen to helium, the second simplest element (something else named after the Greek sun god Helios), releasing a massive amount of energy in the form of heat, light and other kinds of radiation. You remember I said that the size of a star is a balance between the outward push of heat and the inward pull of gravity? Well, this balance stays roughly the same, keeping the star simmering away for several billions of years, until it starts to run out of fuel. What usually happens then is that the star collapses into itself under the unrestrained influence of gravity – at which point all hell breaks loose (if it's possible to imagine anything more hellish than the interior of a star already is).

The life story of a star is too long for astronomers to see more than a tiny snapshot of it. Fortunately, as they scan the skies with their telescopes, astronomers can find a range of stars, each at a different stage of its development: some 'infant' stars caught in the act of being formed from clouds of gas and dust, as our sun was four and a half billion years ago; plenty of 'middle-aged' stars like our sun; and some old and dying stars, which give a foretaste of what will happen to our sun in another few billion years' time. Astronomers have built up a rich 'zoo' of stars, of all different sizes and stages in their life cycles. Each member of the 'zoo' shows what others used to be like, or will be like.

An ordinary star like our sun eventually runs out of hydrogen and, as I've just described, starts 'burning' helium instead (I've put that in quotation marks because it isn't really burning but doing something much hotter). At this stage it is called a 'red giant'. The sun will become a red giant in about five billion years' time, which means it is pretty much in the middle of its life cycle at the moment. Long before then, our poor little planet will have become much too hot to live on. In two billion years the sun will be 15 per cent brighter than it is now, which means that the Earth will be like Venus is today. Nobody could live on Venus: the temperature there is over 400 degrees Celsius. But two billion years is a pretty long time, and humans will almost certainly be extinct long before then, so that there will be nobody left to fry. Or maybe our technology will have advanced to the point where we can actually move the Earth out to a more comfortable orbit. Later, when the helium, too, runs out, the sun will mostly disappear in a cloud of dust and debris, leaving a tiny core called a white dwarf, which will cool and fade.

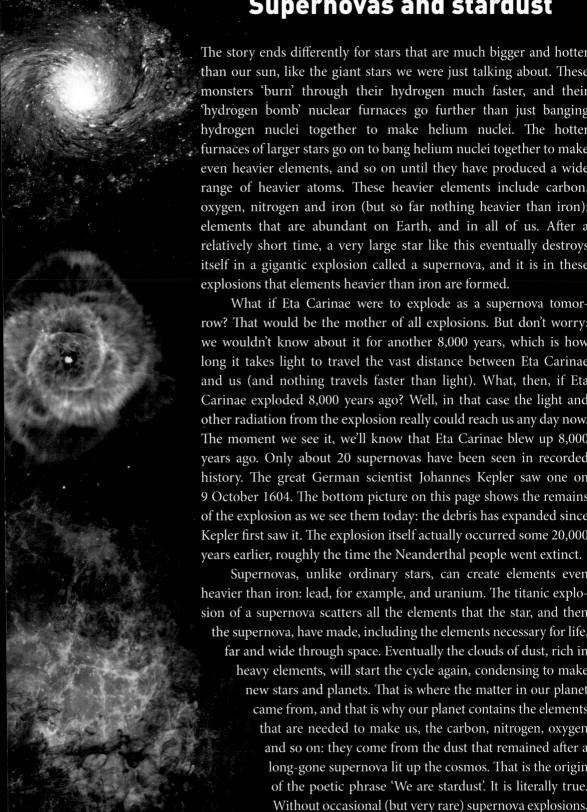

Supernovas and stardust

The story ends differently for stars that are much bigger and hotter than our sun, like the giant stars we were just talking about. These monsters 'burn' through their hydrogen much faster, and their 'hydrogen bomb' nuclear furnaces go further than just banging hydrogen nuclei together to make helium nuclei. The hotter furnaces of larger stars go on to bang helium nuclei together to make even heavier elements, and so on until they have produced a wide range of heavier atoms. These heavier elements include carbon, oxygen, nitrogen and iron (but so far nothing heavier than iron): elements that are abundant on Earth, and in all of us. After a relatively short time, a very large star like this eventually destroys itself in a gigantic explosion called a supernova, and it is in these explosions that elements heavier than iron are formed.

What if Eta Carinae were to explode as a supernova tomorrow? That would be the mother of all explosions. But don't worry: we wouldn't know about it for another 8,000 years, which is how long it takes light to travel the vast distance between Eta Carinae and us (and nothing travels faster than light). What, then, if Eta Carinae exploded 8,000 years ago? Well, in that case the light and other radiation from the explosion really could reach us any day now. The moment we see it, we'll know that Eta Carinae blew up 8,000 years ago. Only about 20 supernovas have been seen in recorded history. The great German scientist Johannes Kepler saw one on 9 October 1604. The bottom picture on this page shows the remains of the explosion as we see them today: the debris has expanded since Kepler first saw it. The explosion itself actually occurred some 20,000 years earlier, roughly the time the Neanderthal people went extinct.

Supernovas, unlike ordinary stars, can create elements even heavier than iron: lead, for example, and uranium. The titanic explosion of a supernova scatters all the elements that the star, and then the supernova, have made, including the elements necessary for life, far and wide through space. Eventually the clouds of dust, rich in heavy elements, will start the cycle again, condensing to make new stars and planets. That is where the matter in our planet came from, and that is why our planet contains the elements that are needed to make us, the carbon, nitrogen, oxygen and so on: they come from the dust that remained after a long-gone supernova lit up the cosmos. That is the origin of the poetic phrase 'We are stardust'. It is literally true. Without occasional (but very rare) supernova explosions, the elements necessary for life would not exist.

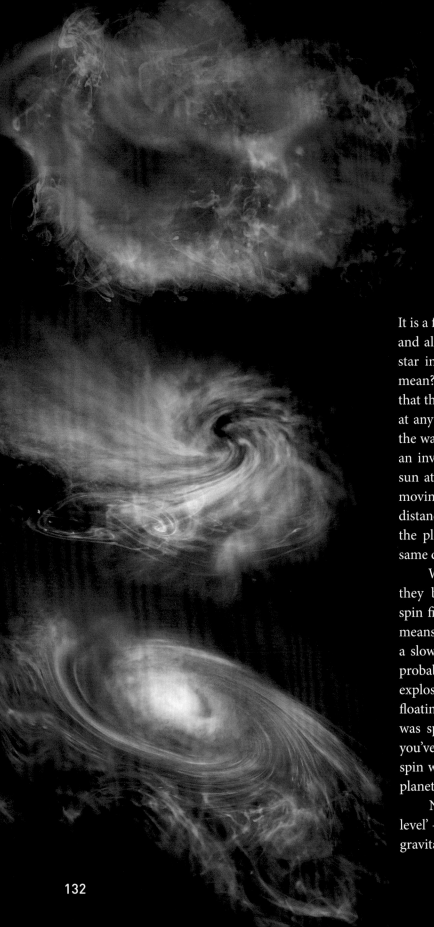

Going round and around

It is a fact we cannot ignore that the Earth and all the sun's other planets orbit their star in the same 'plane'. What does that mean? Theoretically, you might think that the orbit of one planet could be tilted at any angle to any other. But that is not the way things are. It is as though there is an invisible flat disc in the sky, with the sun at the centre, and all the planets are moving on that disc, just at different distances from the centre. What's more, the planets all go round the sun in the same direction.

Why? It is probably because of how they began. Let's take the direction of spin first. The whole solar system, which means the sun and the planets, began as a slowly spinning cloud of gas and dust, probably the leftovers of a supernova explosion. Like almost every other free-floating object in the universe, the cloud was spinning on its own axis. And yes, you've guessed it: the direction of its spin was the same as the direction of the planets now orbiting the sun.

Now, why are all the planets 'on the level' – on that flat 'disc'? For complicated gravitational reasons that I won't go into,

but which scientists understand well, a big spinning cloud of gas and dust out in space tends to form itself into a revolving disc, with a massive lump in the middle. And that is what seems to have happened with our solar system. Dust and gas and small chunks of matter don't stay as gas and dust. Gravitational attraction pulls them towards their neighbours, in the way I described earlier in this chapter. They join forces with those neighbours and form larger lumps of matter. The larger a lump, the greater its gravitational pulling power. So, what happened in our spinning disc was that the larger lumps became even larger, as they sucked in their smaller neighbours.

By far the largest lump became the sun in the centre. Other lumps, large enough to attract smaller lumps to them and far enough from the sun not to be sucked into it, became the planets. Reading from nearest the sun outwards, we now call them Mercury, Venus, Earth, Mars, Jupiter, Saturn, Uranus and Neptune. Old lists would put Pluto after Neptune, but nowadays it is regarded as too small to count as a planet.

Under different circumstances another planet could have formed too, between the orbits of Mars and Jupiter. But the small bits that could otherwise have joined together to make this extra planet were prevented from doing so, probably by the brooding gravitational presence of Jupiter, and they have remained as an orbiting ring of debris called the asteroid belt. These asteroids swarm in a ring between the orbits of Mars and Jupiter, which i where the extra planet would have been if they ha managed to get together. The famous rings aroun the planet Saturn are there for a similar reason They could have condensed together to make an other moon (Saturn already has 62 moons, so thi would have been the 63rd), but they actually staye separate as a ring of rocks and dust. In the asteroid belt – the sun's equivalent of Saturn's rings – som of the bits of debris are large enough to be calle planetesimals (sort of 'not quite planets'). The larg est of them, called Ceres, is nearly 1,000 kilometre across, large enough to be roughly spherical lik a planet, but most of them are just mis-shape rocks and bits of dust. They collide with eac other from time to time, like billiard ball and sometimes one of them ge kicked out of the asteroid belt ar may even come close to another plan such as Earth.

We see them, quite commonly, burr ing in the upper atmosphere as 'shootir stars' or 'meteors'.

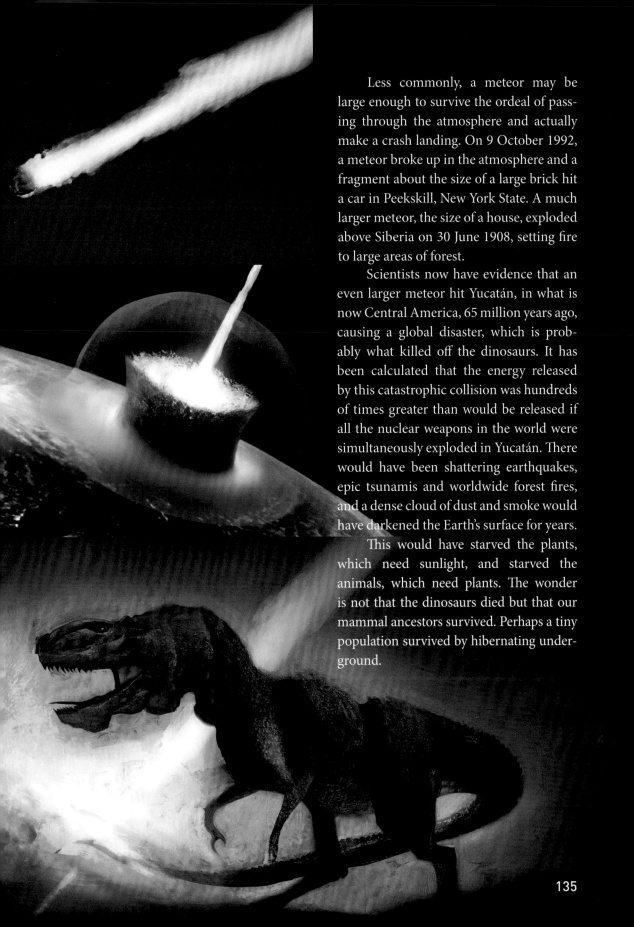

Less commonly, a meteor may be large enough to survive the ordeal of passing through the atmosphere and actually make a crash landing. On 9 October 1992, a meteor broke up in the atmosphere and a fragment about the size of a large brick hit a car in Peekskill, New York State. A much larger meteor, the size of a house, exploded above Siberia on 30 June 1908, setting fire to large areas of forest.

Scientists now have evidence that an even larger meteor hit Yucatán, in what is now Central America, 65 million years ago, causing a global disaster, which is probably what killed off the dinosaurs. It has been calculated that the energy released by this catastrophic collision was hundreds of times greater than would be released if all the nuclear weapons in the world were simultaneously exploded in Yucatán. There would have been shattering earthquakes, epic tsunamis and worldwide forest fires, and a dense cloud of dust and smoke would have darkened the Earth's surface for years.

This would have starved the plants, which need sunlight, and starved the animals, which need plants. The wonder is not that the dinosaurs died but that our mammal ancestors survived. Perhaps a tiny population survived by hibernating underground.

Light of our lives

I want to end this chapter by talking about the importance of the sun for life. We don't know whether there is life elsewhere in the universe (I'll discuss that question in a later chapter), but we do know that, if there is life out there, it is almost certainly near a star. We can also say that, if it is anything like our kind of life, at least, it will probably be on a planet about the same apparent distance from its star as we are from our sun. By 'apparent distance' I mean distance as perceived by the life form itself. The absolute distance could be very much greater, as we saw in the example of the super-giant star R136a1. But if the apparent distance were the same, their sun would look about the same size to them as ours does to us, which would mean that the amount of heat and light received from it would be about the same.

Why does life have to be close to a star? Because all life needs energy, and the obvious source of energy is starlight. On Earth, plants gather sunlight and make its energy available to all other living creatures. Plants could be said to feed off sunlight. They need other things too, such as carbon dioxide from the air, and water and minerals from the ground. But they get their energy from sunlight, and they use it to make sugars, which are a kind of fuel

that drives everything else that they need to do.

You can't make sugar without energy. And once you have sugar, you can then 'burn' it to get the energy back out again – though you never get all of the energy back; there is always some lost in the process. And when we say 'burn', that doesn't mean it goes up in smoke. Literally burning it is only one way to release the energy in a fuel. There are more controlled ways to let the energy trickle out, slowly and usefully.

You can think of a green leaf as a low, spread-out factory whose entire flat roof is one great solar panel, trapping sunlight and using it to drive the wheels of the assembly lines under the roof. That is why leaves are thin and flat – to give them a large surface area for sunlight to fall on. The end product of the factory is sugars of various kinds. These are then piped through the veins in the leaf to the rest of the plant, where they are used to make other things, like starch, which is a more convenient way to store energy than sugar. Eventually, the energy is released from the starch or sugar to make all the other parts of the plant.

When plants are eaten by herbivores (which means just that: 'plant-eaters'), such as antelopes or rabbits, the energy is passed to the herbivores – and again, some of it is lost in the process. The herbivores use it

Eating

Digestion

to build up their bodies and fuel their muscles as they go about their business. Their business includes, of course, grazing or browsing on lots more plants. The energy that powers the muscles of the herbivores as they walk and munch and fight and mate comes ultimately from the sun, via plants.

Then other animals – meat-eaters or 'carnivores' – come along and eat the herbivores. The energy is passed on yet again (and yet again some of it is lost in the transition), and it powers the muscles of the carnivores as they go about their business. In this case, their business includes hunting down yet more herbivores to eat, as well as all the other things they do, like mating and fighting and climbing trees and, in the case of mammals, making milk for their babies. Still, it is the sun that ultimately provides the energy, even though by now that energy has reached them by a very indirect route. And at every stage of that indirect route, a good fraction of the energy is lost – lost as heat, which contributes to the useless task of heating up the rest of the universe.

Other animals, parasites, feed on the living bodies of both herbivores and carnivores. Once again, the energy that powers the parasites comes ultimately from the sun, and once again not all of it is used because some of it is wasted as heat.

Finally, when anything dies, whether plant or herbivore or carnivore or parasite, it may be eaten by scavengers like burying beetles, or it may decay – eaten by bacteria and fungi, which are just a different kind of scavenger. Yet again, the energy from the sun is handed on, and yet again some of it leaks away as heat. That's why compost heaps are hot. All the heat in a compost heap comes ultimately from the sun, trapped by leafy solar panels the year before. There are fascinating Australasian birds called megapodes that use the heat of a compost heap to incubate their eggs. Unlike other birds, which sit on their eggs and warm them with their body heat, megapodes build a big compost heap in which they lay their eggs. They regulate the temperature of the heap by piling more compost on the top to make it hotter, or removing compost to make it cooler. But all birds ultimately use solar energy to incubate their eggs, whether through their body heat or through a compost heap.

Sometimes plants are not eaten but sink into peat bogs. Over centuries, they become compressed into layers of peat by new layers added above them. People in western Ireland or the Scottish isles dig up the peat and cut it into brick-sized chunks, which they burn as fuel, to keep their houses warm in winter. Once again, it is trapped sunlight – in this case trapped centuries earlier – whose energy is being released in the fires and cooking ranges of Galway and the Hebrides.

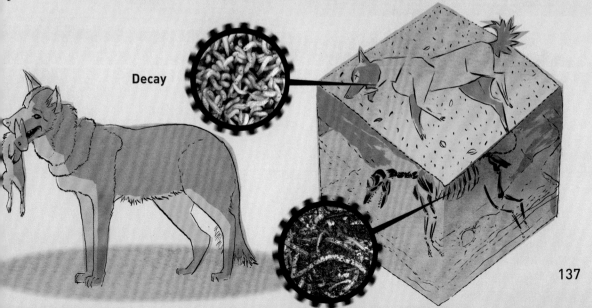

Decay

Under the right conditions, and over millions of years, peat can become compacted and transformed, so that it eventually becomes coal. Weight for weight, coal is a more efficient fuel than peat and burns at a much higher temperature, and it was coal fires and furnaces that powered the industrial revolution of the eighteenth and nineteenth centuries.

The intense heat of a steel mill or a blast furnace, the glowing fireboxes that sent the Victorians' steam engines thundering along iron rails or their ships pounding through the sea: all that heat came originally from the sun, via the green leaves of plants that lived 300 million years ago.

Some of the 'dark Satanic mills' of the industrial revolution were driven by steam power, but many of the earlier cotton mills were powered by water wheels. The mill was built near a fast-running river, which was ducted to flow over a wheel. This water wheel turned a great axle or drive shaft, which ran the length of the factory. All along the drive shaft, belts and cogwheels drove the various spinning machines and carding machines and looms. Even those machines were ultimately driven by the sun. Here's how.

The water wheels were driven by water, being pulled downhill by gravity. But that works only because there is a continuous supply of water up on the high ground, from where it can run downhill. That water is supplied in the form of rain, from clouds, falling on the hills and mountains. And the clouds get their water through the evaporation of seas, lakes, rivers and puddles on Earth. Evaporation requires energy, and that energy comes from the sun. So ultimately the energy that drove the water wheels that turned the belts and cogwheels of the spinning machines and looms all came from the sun.

Later cotton mills were driven by coal-fired steam engines – again using energy ultimately from the sun. But before they switched to steam entirely the factories went through an intermediate stage. They kept the great water wheel to drive the looms and shuttles, but used a steam engine to pump water up into a tank, from which it flowed down over the water wheel, only to be pumped back up again. So, whether the water is raised by the sun into the clouds, or whether it is raised by a coal-fired steam engine into a tank, the energy still comes from the sun in the first place. The difference is that the steam engine is driven by sunlight collected by plants millions of years ago and stored underground in coal, whereas the water wheel on a river is driven by sunlight from

It wouldn't do us any good if we literally burned our sugar and other food fuels by setting fire to them! Burning is a wasteful and destructive way to recover the sun's stored energy. What happens in our cells is so slow and carefully regulated that it is like water trickling down a hill and driving a series of water wheels. The sun-powered chemical reaction that goes on in green leaves to make sugar is doing the equivalent of pumping water uphill. The chemical reactions in animal and plant cells that use energy – to drive muscles, for example – get the energy in carefully controlled stages, step by step. The high-energy fuels, sugars or whatever they are, are coaxed into releasing their energy in stages, down through a cascade of chemical reactions, each one feeding into the next, like a stream tumbling down a series of small waterfalls, turning one small water wheel after another.

Whatever the details, all the water wheels and cogs and drive shafts of life are ultimately powered by the sun. Perhaps those ancient peoples would have worshipped the sun even more devotedly if they had realized just how much all life depended on it. What I now wonder is how many other stars drive engines of life on their own orbiting planets. But that must wait for a later chapter.

only a few weeks ago and stored in the form of the water up at the top of the hills. This kind of 'stored sunlight' is called potential energy, because the water has the potential – the power within it – to do work as it flows downhill.

This gives us a nice way to understand how life is powered by the sun. When plants use sunlight to make sugar, it is like pumping water uphill, or into a tank on a factory roof. When plants (or the herbivores that eat the plants, or the carnivores that eat the herbivores) use the sugar (or the starch that's made from the sugar, or the meat that's made from the starch), we can think of the sugar as being burned: slow-burned to drive muscles, for instance, just as coal is fast-burned to make steam to propel a drive shaft in a factory.

What *is* *a*

7

THE EPIC OF Gilgamesh is one of the oldest stories ever written. Older than the legends of the Greeks or the Jews, it is the ancient heroic myth of the Sumerian civilization, which flourished in Mesopotamia (now Iraq) between 5,000 and 6,000 years ago. Gilgamesh was the great hero king of Sumerian myth – a bit like King Arthur in British legends, in that nobody knows whether he actually existed, but lots of stories were told about him. Like the Greek hero Odysseus (Ulysses) and the Arabian hero Sinbad the Sailor, Gilgamesh went on epic travels, and he met many strange things and people on his journeys. One of them was an old man (a very, very old man, centuries old) called Utnapashtim, who told Gilgamesh a strange story about himself. Well, it seemed strange to Gilgamesh, but it may not seem so strange to you because you have probably heard a similar story . . . about another old man with a different name.

rainbow?

Utnapashtim told Gilgamesh of an occasion, many centuries earlier, when the gods were angry with human-kind because we made so much noise they couldn't sleep.

The chief god, Enlil, suggested that they should send a great flood to destroy everybody, so the gods could get a good night's rest. But the water god, Ea, decided to warn Utnapashtim. Ea told Utnapashtim to tear down his house and build a boat.

It would have to be a very big boat, because Utnapashtim was to take into it 'the seed of all living creatures'.

Utnapashtim built the boat just in time, before it rained for six days and six nights without stopping. The flood that followed drowned everybody and everything that was not safely inside the boat. On the seventh day the wind dropped and the waters grew calm and flat.

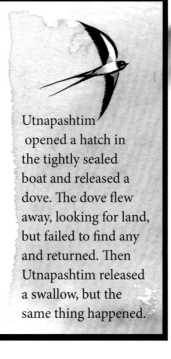

Utnapashtim opened a hatch in the tightly sealed boat and released a dove. The dove flew away, looking for land, but failed to find any and returned. Then Utnapashtim released a swallow, but the same thing happened.

Finally Utnapashtim released a raven. The raven didn't come back, which suggested to Utnapashtim that there was dry land somewhere and the raven had found it.

Eventually the boat came to rest on a mountaintop poking out of the water. Another god, Ishtar, created the first rainbow, as a token of the gods' promise to send no more terrible floods. So that is how the rainbow came into being, according to the ancient legend of the Sumerians.

Well, I said the story would be familiar. All children reared in Christian, Jewish or Islamic countries will immediately recognize that it is the same as the more recent story of Noah's Ark, with one or two minor differences. The name of the boat-builder changes from Utnapashtim to Noah. The many gods of the older legend turn into the one god of the Jewish story. The 'seed of all living creatures' becomes spelled out as 'every living thing of all flesh, two of every sort' – or, as the song has it, 'the animals went in two by two' – and the Epic of Gilgamesh surely meant something similar. In fact, it is obvious that the Jewish story of Noah is nothing more than a retelling of the older legend of Utnapashtim. It was a folk tale that got passed around, and it travelled down the centuries. We often find that seemingly ancient legends have come from even older legends, usually with some names or other details changed. And this one, in both versions, ends with the rainbow.

In both the Epic of Gilgamesh and the Book of Genesis, the rainbow is an important part of the myth. Genesis specifies that it was actually God's bow, which he put up in the sky as a token of his promise to Noah and his descendants.

There is one more difference between the Noah story and the earlier Sumerian tale of Utnapashtim. In the Noah version, the reason for God's discontent with humans was that we were all incurably wicked. In the Sumerian story, humanity's crime was, you might think, less serious. We simply made so much noise the gods couldn't get to sleep! I think it's funny. And the theme of noisy humans keeping the gods awake crops up, quite independently, in the legend of the Chumash people of Santa Cruz Island, off the coast of California.

The Chumash people believed that they were created on their island (it obviously wasn't called Santa Cruz then, because that is a Spanish name) from the seeds of a magic plant by the Earth goddess Hutash, who was married to the Sky Snake (what we know as the Milky Way, which you can see on a really dark night in the country,

but not if you live in a town where there is too much light pollution). The people of the island became very numerous, and, just as in the Epic of Gilgamesh, too noisy for the goddess Hutash's comfort. The racket kept her awake at night. But instead of killing them all, like the Sumerian and Jewish gods, Hutash was kinder. She decided that some of them must move off Santa Cruz, onto the mainland where she wouldn't be able to hear them. So she made a bridge for them to cross by. And the bridge was . . . yes, the rainbow!

This myth has a strange ending.

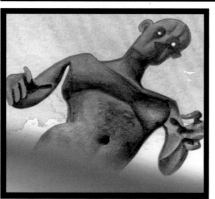

As the people were crossing over the rainbow bridge, some of the noisy ones looked down – and they were so frightened by the drop that they got dizzy.

They fell off the rainbow into the sea, where they turned into dolphins.

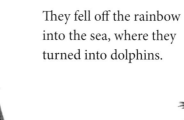

The idea of the rainbow as a bridge crops up in other mythologies, too. In old Norse (Viking) myths, rainbows were seen as fragile bridges used by the gods to travel from the sky world to Earth.

Many peoples, for example in Persia, west Africa, Malaysia, Australia and the Americas, have seen the rainbow as a large snake which soars out of the ground to drink the rain.

How do all these legends start, I wonder? Who makes them up, and why do some people eventually come to believe these things really happened? These questions are fascinating and not easy to answer. But there's one question we can answer: what is a rainbow *really*?

The real magic of the rainbow

WHEN I WAS about ten, I was taken to London to see a children's play called *Where the Rainbow Ends*. You almost certainly won't have seen it because it is too unfashionably patriotic for modern theatres to perform. It is all about how exceptionally special it is to be English, and at the climax of the adventure the children are rescued by St George, the patron saint of England (not Britain, for Scotland, Wales and Ireland have their own patron saints). But what I most vividly remember is not St George but the rainbow itself. The children actually went to the place where the rainbow planted its foot, and we saw them walking about in the middle of the rainbow where it hit the ground. It was cleverly staged, with coloured spotlights beaming down through swirling mist, and the children stumbled about in a spellbound daze. I think it was at about this moment that the shining-armoured, silver-helmeted St George appeared, and we children gasped at the scene as the children on the stage shouted:

'St George! St George! St George!'

But it was the rainbow itself that seized my imagination. Never mind St George: how wonderful it must be to stand right in the foot of a giant rainbow!

You can see where the author of the play got the idea. A rainbow really does look like a proper object, hanging out there, perhaps a few miles away. It seems to have its left foot planted, say, in a wheat field and its right foot (if you are lucky enough to see a complete rainbow) on a hilltop. You feel you ought to be able to go straight to it and stand right where the rainbow steps on the ground, like the children in the play. All the myths I have described to you have the same idea. The rainbow is seen as a definite thing, in a definite place, a definite distance away.

Well, you'll probably have worked out that it isn't really like that! First, if you try to approach the rainbow, no matter how fast you run, you'll never get there: the rainbow will run away from you until it fades away altogether. You can't catch it. But it isn't really running away because it isn't really in a particular place at all, ever. It's an illusion – but a fascinating illusion, and understanding it leads on to all sorts of interesting things, some of which we'll come to in the next chapter.

What light is made of

First, we need to understand about something called the spectrum. It was discovered in the time of King Charles II – that's about 350 years ago – by Isaac Newton, who may well have been the greatest scientist ever (he discovered lots of other things besides the spectrum, as we saw in the chapter on night and day). Newton discovered that white light is really a mixture of all the different colours. To a scientist, that's what white *means*.

How did Newton find this out? He set up an experiment. First he blacked out his room so that no light could get in, and then he opened a narrow chink in the curtain, so that a pencil-thin beam of white sunlight came in. He then let the beam of light pass through a prism, which is a sort of triangular chunk of glass.

What a prism does is splay the narrow white beam out; but the splayed-out beam that emerges from the prism is no longer white. It is multi-coloured like a rainbow, and Newton gave a name to the rainbow he made: the spectrum. Here's how it works.

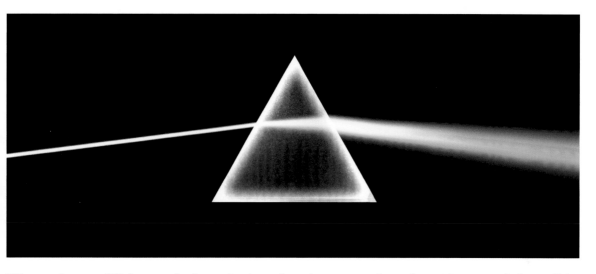

When a beam of light travels through air and hits glass, it gets bent. The bending is called refraction. Refraction doesn't have to be caused by glass: water does the trick too, and that will be important when we come back to the rainbow. It is refraction that makes an oar look bent when

you stick it in the river. So, light is bent when it hits glass or water. But now here's the point. The *angle* of the bend is slightly different depending on what colour the light is. Red light bends at a shallower angle than blue light. So, if white light really is a mixture of coloured lights, as Newton guessed, what's going to happen when you bend white light through a prism? The blue light is going to bend further than the red light, so they will be separated from each other when they emerge from the other side of the prism. And the yellow and green lights will come out in between. The result is Newton's spectrum: all the colours of the rainbow, arranged in the correct rainbow order – red, orange, yellow, green, blue, violet.

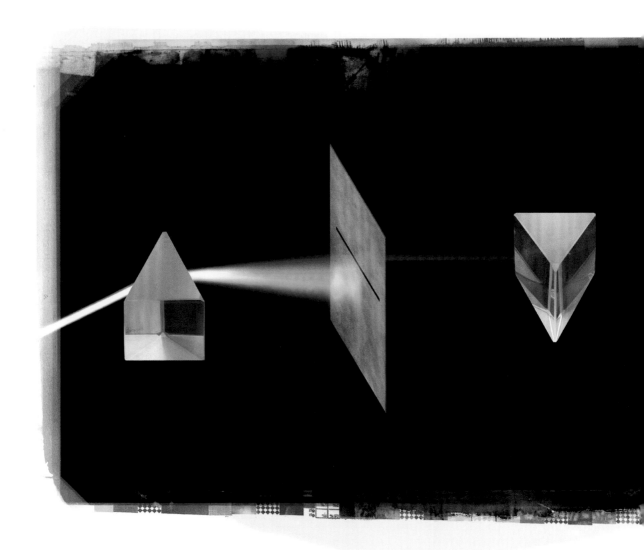

Newton wasn't the first person to make a rainbow with a prism. Other people had already got the same result. But many of them thought the prism somehow 'coloured' the white light, like adding a dye. Newton's idea was quite different. He thought that white light was a mixture of all the colours, and the prism was just separating them from each other. He was right, and he proved it with a pair of neat experiments. First, he took his prism, as before, and stuck a narrow slit in the way of the coloured beams coming out of it, so that only one of them, say the red beam, passed through the slit. Then he put another prism in the path of this narrow beam of red light. The second prism bent the light, as usual. But what came out of it was only red light. No extra colours were added, as they would have been if what prisms did was add colour like a dye. The result Newton got was exactly what he expected, supporting his theory that white light is a mixture of light of all colours.

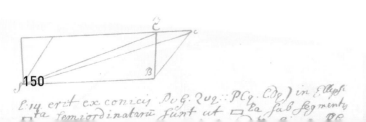

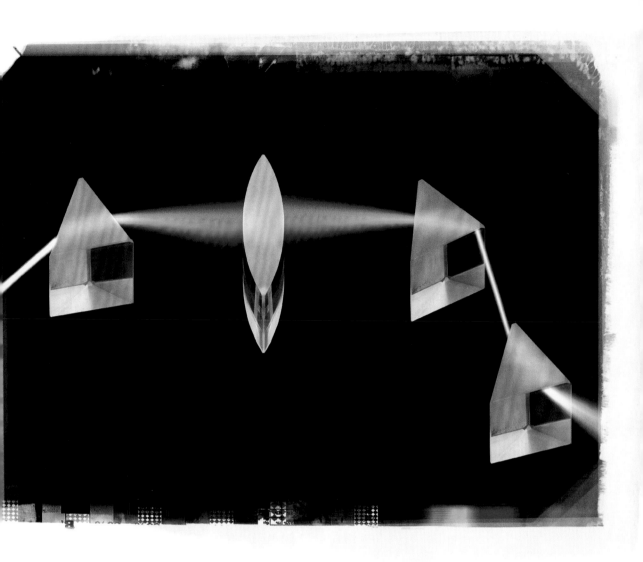

The second experiment was more ingenious still, using three prisms. It was called Newton's Experimentum Crucis, which is Latin for 'critical experiment' – or, as we might say, 'experiment that really clinches the argument'.

On the left of the picture above you see white light coming through a slit in Newton's curtain and passing through the first prism, which spreads it out into all the colours of the rainbow. The spread-out rainbow colours then pass through a lens, which brings them all together before they pass through the second of Newton's prisms. This second prism had the effect of merging the rainbow colours back into white light again. That already neatly proved Newton's point. But just to make quite sure, he then passed the beam of white light through a third prism, which splayed the colours out into a rainbow again! As neat a demonstration as you could wish for, proving that white light is indeed a mixture of all the colours.

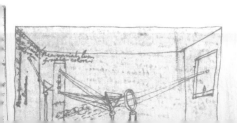

Prisms are all very well, but when you see a rainbow in the sky, there isn't a great big prism hanging up there. No, but there are millions of raindrops. So, does each raindrop act as a tiny prism? It is a bit like that, but not quite.

If you want to see a rainbow you have to have the sun *behind* you when you look at a rainstorm. Each raindrop is more like a little ball than a prism, and light behaves differently when it hits a ball from how it behaves when it hits a prism. The difference is that the far side of a raindrop acts as a tiny mirror. And that is why you need the sun behind you if you want to see a rainbow. The light from the sun turns a somersault inside every raindrop and is reflected backwards and downwards, where it hits your eyes.

Here's how it works. You are standing with the sun *behind* and above you, looking at a distant shower of rain. The sunlight hits a single raindrop (of course it hits lots of other raindrops too, but wait, we're coming to that). Let's call our one particular raindrop A. The beam of white light hits A on its upper near surface, where it is bent, just as it was on the near surface of Newton's prism. And of course the red light bends less than the blue, so the spectrum is already sorting itself out. Now all the coloured beams travel through the raindrop until they hit the far side. Instead of passing through into the air, they are reflected back towards the near side of the raindrop, this time the lower part of the near side. And as they pass through the near side of the raindrop, they are again bent. Again the red light bends less than the blue.

So, as the sunbeam leaves the raindrop, it has been splayed out into a proper little spectrum. The separated coloured beams, having doubled back around the inside of the raindrop, are now hurtling back in the general direction of where you are standing. If your eye happens to be in the path of one of those beams, say the green one, you'll see pure green light. Somebody shorter than you might see the red beam coming from A. And somebody taller than you might see the blue beam from A.

Nobody sees the full spectrum from any one raindrop. Each of you sees only one pure colour. Yet all of you say you see a rainbow, with all the colours. How come? Well, so far, we have only been talking about one raindrop, called *A*. There are millions of other raindrops, and they are all behaving in the same kind of way. While you are looking at *A*'s red beam, there is another raindrop called *B*, which is lower than *A*. You don't see *B*'s red beam because it hits you in the stomach. But *B*'s blue beam is in exactly the right place to hit you in the eye. And there are other raindrops lower than *A* but higher than *B*, whose red and blue beams miss your eye but whose yellow or green beams hit your eye. So lots of raindrops together add up to a complete spectrum, in a line, up and down.

But a line up and down is not a rainbow. Where does the rest of the rainbow come from? Don't forget that there are other raindrops, stretching from one side of the rain shower to the other and at all heights. And of course they fill in the rest of the rainbow for you. Every rainbow you see, by the way, is trying to be a complete circle, with your eye at the centre of it – like the complete circular rainbow you sometimes see when you water the garden with a hose and the sun shines through the spray. The only reason we don't usually see the whole circle is that the ground gets in the way.

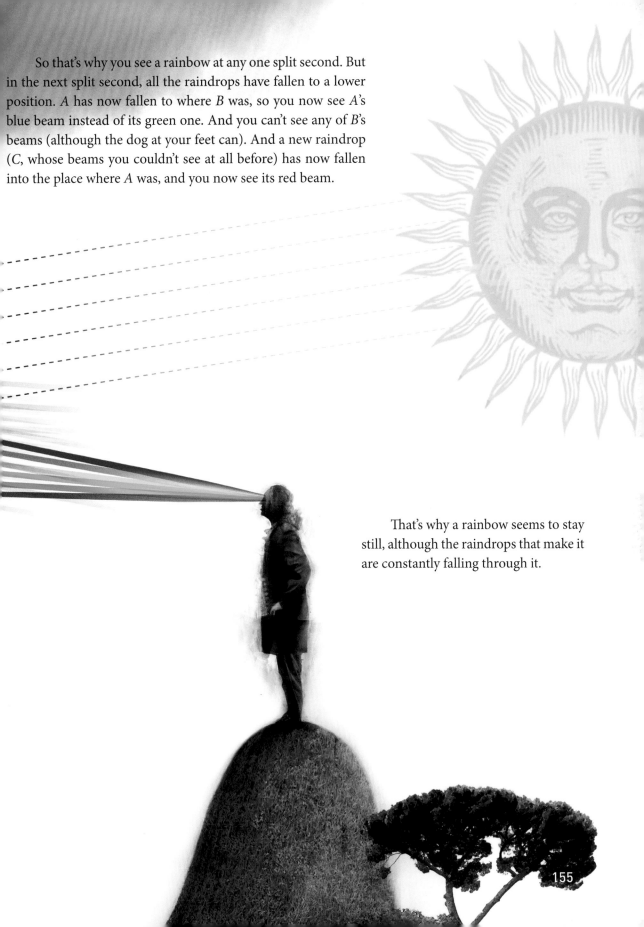

So that's why you see a rainbow at any one split second. But in the next split second, all the raindrops have fallen to a lower position. *A* has now fallen to where *B* was, so you now see *A*'s blue beam instead of its green one. And you can't see any of *B*'s beams (although the dog at your feet can). And a new raindrop (*C*, whose beams you couldn't see at all before) has now fallen into the place where *A* was, and you now see its red beam.

That's why a rainbow seems to stay still, although the raindrops that make it are constantly falling through it.

On the right 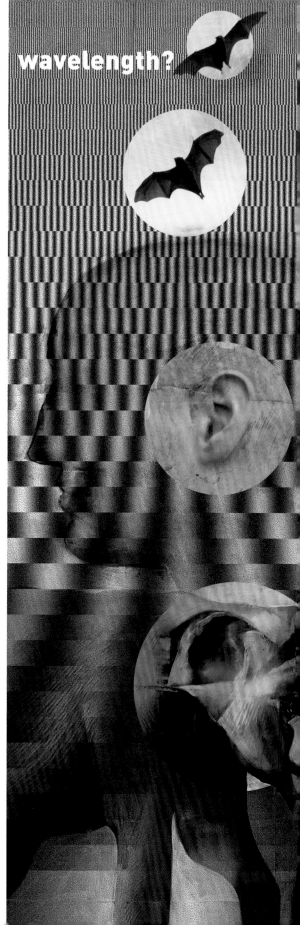wavelength?

Let's now look at what the spectrum – the ordered range of colours from red through orange, yellow, green and blue to violet – really is. What is it about red light that makes it bend at a shallower angle than blue light?

Light can be thought of as vibrations: waves. Just as sound is vibrations in the air, light consists of what are called electromagnetic vibrations. I won't try to explain what electromagnetic vibrations are because it takes too long (and I'm not sure that I entirely understand it myself). The point here is that although light is very different from sound, we can talk about high-frequency (short-wavelength) and low-frequency (long-wavelength) vibrations in light, just as we can for sound. High-pitched sound – treble or soprano – means high-frequency, or short-wavelength, vibrations. Low-frequency, or long-wavelength, sounds are deep, bass sounds. The equivalent for light is that red (long wavelength) is the bass, yellow the baritone, green the tenor, blue the alto and violet (short wavelength) the treble.

There are sounds that are too high-pitched for us to hear. They are called ultrasound; bats can hear them and use the echoes for finding their way around. There are also sounds that are too low for us to hear. They are called infrasound; elephants, whales and some other animals use these deep rumbles for keeping in touch with each other. The deepest bass notes on a big cathedral organ are almost too low to hear: you seem to 'feel' them fluttering your whole body. The range of sounds that we humans can hear is a band of frequencies in the middle, between ultrasound, which is too high for us (but not bats) to hear, and infrasound, which is too low for us (but not elephants) to hear.

And the same is true of light. The colour equivalent of ultrasound bat squeaks is ultraviolet, which means 'beyond violet'. Although we can't see ultraviolet light, insects can. There are some flowers that have stripes or other patterns for luring insects in to pollinate them, patterns that can only be seen in the ultraviolet range of wavelengths. Insect eyes can see them, but we need instruments to 'translate' the patterns into the visible part of the spectrum. The evening primrose flower on the right looks yellow to us, with no pattern, no stripes. But if you photograph it in ultraviolet light you suddenly see a starburst of stripes. The pattern in the lower picture is not really white but ultraviolet. Since we can't see ultraviolet, we have to represent the pattern in some colour that we can see, and the person who made the picture decided to use black and white. He could have chosen blue or any colour.

The spectrum goes into higher and higher frequencies, far beyond ultraviolet, far beyond what even insects can see. X-rays could be thought of as 'light' of even higher 'pitch' than ultraviolet. And gamma rays are even higher still.

At the other end of the spectrum, insects can't see red, but we can. Beyond red is 'infrared', which we can't see, although we can feel it as heat (and some snakes are especially sensitive to it, using it to detect their prey). I suppose a bee might call red 'infra-orange'. Deeper 'bass notes' than infrared are microwaves, which you use to cook things. And even deeper bass (longer wavelength) are radio waves.

INFRARED

VISIBLE LIGHT

ULTRAVIOLET

X-RAYS

MICROWAVES

RADIO WAVES

What is a bit surprising is that the light we humans can actually see – the spectrum or 'rainbow' of visible colours between the slightly 'higher-pitched' violet and the slightly 'lower-pitched' red – is a tiny band in the middle of a huge spectrum ranging from gamma rays at the high-pitched end to radio waves at the low-pitched end. Almost the whole of the spectrum is invisible to our eyes.

The sun and the stars are pumping out electromagnetic rays at a full range of frequencies or 'pitches', all the way from radio waves at the 'bass' end to cosmic rays at the 'treble' end. Although we can't see outside the tiny band of visible light, from red to violet, we have instruments that can detect these invisible rays. The picture of the supernova in Chapter 6 was taken using X-rays from the supernova. The colours in the picture are false colours, like the false white used to show the pattern on the evening primrose flower. In the supernova picture, false colours are used to designate different wavelengths of X-rays. Scientists called radio astronomers take 'photographs' of stars using radio waves rather than light waves or X-rays. The instrument they use is called a radio telescope. Other scientists take photographs of the sky at the other end of the spectrum, in the X-ray band. We learn different things about the stars and about the universe by using different parts of the spectrum. The fact that our eyes can see through only a tiny slit in the middle of the vast spectrum, that we can see only a slender band in the huge range of rays that scientific instruments can see, is a lovely illustration of the power of science to excite our imagination: a lovely example of the magic of the real.

In the next chapter we shall learn something even more wonderful about the rainbow. Splitting the light from a distant star into a spectrum can tell us not only what the star is made of but also how old it is. And it is evidence of this kind – rainbow evidence – that enables us to work out how old the universe is: when did it all begin? That may sound unlikely, but all will be revealed in the next chapter.

8 WHEN AND HOW DID EVERYTHING BEGIN?

LET'S START with an African myth from a Bantu tribe, the Boshongo of the Congo. In the beginning there was no land, just watery darkness, and also – importantly – the god Bumba. Bumba got a stomach-ache and vomited up the sun. Light from the sun dispelled the darkness, and heat from the sun dried up some of the water, leaving land. Bumba's stomach-ache still hadn't gone, though, so he then sicked up the moon, the stars, animals and people.

Many Chinese origin myths involve a character called Pan Gu, sometimes depicted as a giant hairy man with a dog's head. Here's one of the Pan Gu myths. In the beginning there was no clear distinction between Heaven and Earth: it was all one gooey mess surrounding a big black egg. Curled up inside the egg was Pan Gu. Pan Gu slept inside the egg for 18,000 years. When he finally awoke he wanted to escape, so he picked up his axe and hewed his way out. Some of the contents of the egg were heavy and sank to become the Earth. Some of them were light and floated up to become the sky. The Earth and the sky then swelled at a rate of (the equivalent of) 3 metres a day for another 18,000 years.

Some versions of the story have Pan Gu pushing the sky and the Earth apart, after which he was so exhausted that he died. Various bits of him then became the universe that we know. His breath became the wind, his voice became thunder; his two eyes became the moon and the sun, his muscles farmland and his veins roads. His sweat became rain, and his hairs became stars. Humans are descended from the fleas and lice that once lived on his body.

By the way, the story of Pan Gu pushing the sky and the Earth apart is rather like the (probably unrelated) Greek myth of Atlas, who also held up the sky (although, weirdly, pictures and statues usually show him carrying the whole Earth on his shoulders).

Now here is one of many origin myths from India. Before the beginning of time there was a great dark ocean of nothingness, with a giant snake coiled up on the surface. Sleeping in the coils of the snake was Lord Vishnu. Eventually Lord Vishnu was awakened by a deep humming sound from the bottom of the ocean of nothingness, and a lotus plant grew out of his navel. In the middle of the lotus flower sat Brahma, Vishnu's servant. Vishnu commanded Brahma to create the world. So Brahma did just that. No problem! And all living creatures too, while he was about it. Easy!

What I find a little disappointing about all these origin myths is that they begin by assuming the existence of some kind of living creature before the universe itself came into being – Bumba or Brahma or Pan Gu, or Unkulukulu (the Zulu creator) or Abassie (Nigeria) or 'Old Man in the Sky' (Salish, a tribe of native Americans from Canada). Wouldn't you think that a universe of some kind would have to come first, to provide a place for the creative spirit to go to work? None of the myths gives any explanation for how the creator of the universe himself (and it usually is a he) came into existence.

So they don't get us very far. Let's turn instead to what we know of the true story of how the universe began.

163

HOW DID EVERYTHING BEGIN
EVERYTHING
BEGIN
REALLY?

DO YOU remember from Chapter 1 that scientists work by setting up 'models' of how the real world might be? They then test each model by using it to make predictions of things that we ought to see – or measurements that we ought to be able to make – if the model were correct. In the middle of the twentieth century there were two competing models of how the universe came into being, called the 'steady state' model and the 'big bang' model. The steady state model was very elegant, but eventually turned out to be wrong – that is, predictions based on it were shown to be false. According to the steady state model, there never was a beginning: the universe had always existed in pretty much its present form. The big bang model, on the other hand, suggested that the universe began at a definite moment in time, in a strange kind of explosion. The predictions made on the basis of the big bang model keep turning out to be right, and so it has now been generally accepted by most scientists.

According to the modern version of the big bang model, the entire observable universe exploded into existence between 13 and 14 billion years ago. Why do I say 'observable'? The 'observable universe' means everything for which we have any evidence at all. It is possible that there are other universes that are inaccessible to all our senses and instruments. Some scientists speculate, perhaps fancifully, that there may be a 'multiverse': a bubbling 'foam' of universes, of which our universe is only one 'bubble'. Or it may be that the observable universe – the universe in which we live, and the only universe for which we have direct evidence – is the only universe there is. Either way, in this chapter I am limiting myself to the observable universe. The observable universe seems to have begun in the big bang, and this remarkable event happened just under 14 billion years ago.

Some scientists will tell you that time itself began in the big bang, and we should no more ask what happened before the big bang than we should ask what is north of the North Pole. You don't understand that? Nor do I. But I do understand, sort of, the evidence that the big bang happened, and when. That is what this chapter is about.

First, I need to explain what a galaxy is. We've already seen, in our analogy with footballs in Chapter 6, that the stars are spaced out at incredibly huge distances from one another compared with the planets orbiting our sun. But, vastly spaced out as they are, the stars are still actually clustered together into groups; and these groups are called galaxies. Here's a picture of four galaxies:

Each galaxy shows up as a white swirling pattern that is actually made up of billions of stars, and also clouds of dust and gas.

Sun

Our sun is just one of the stars that make up the particular galaxy called the Milky Way. It is called that because on dark nights we get an end-on view of part of it. We see it as a mysterious streak or path of white across the sky, which you might mistake for a long, wispy cloud until you realize what it really is – and when you do, the thought should strike you dumb with awe. Since we are in the Milky Way galaxy, we can never see it in its full glory, but above is an artist's impression of it as if seen from outside, with our own position marked. It is marked as 'sun' because on this scale there is no noticeable distance between the sun and any of its planets.

And now here is a picture (right) – not an artist's impression but an actual photograph taken through a telescope – of hundreds of galaxies, each one a huge cluster of billions of stars, just like our Milky Way. I can't help being amazed every time I look at this that each one of those little smudges of light is an entire galaxy, comparable to the Milky Way. But that is the bare fact. The universe – our observable universe – is a very big place.

The next important point is this. It is possible to measure how far away from us each galaxy is. How? How, for that matter, do we know how far away anything in the universe is? For nearby stars the best method uses something called 'parallax'. Hold your finger up in front of your face and look at it with your left eye closed. Now open your left eye and close your right. Keep switching eyes, and you'll notice that the apparent position of your finger hops from side to side. That is because of the difference between the viewpoints of your two eyes. Move your finger nearer, and the hops will become greater. Move your finger further away and the hops become smaller. All you need to know is how far apart your eyes are, and you can calculate the distance from eyes to finger by the size of the hops. That is the parallax method of estimating distances.

Now, instead of looking at your finger, look at a star out in the night sky, switching from eye to eye. The star won't hop at all. It is much too far away. In order to make a star 'hop' from side to side, your eyes would need to be millions of miles apart! How can we achieve the same effect as switching eyes millions of miles apart? We can make use of the fact that the Earth's orbit around the sun has a diameter of 186 million miles. We measure the position of a nearby star, against a background of other stars. Then, six months later, when the Earth is 186 million miles away at the opposite side of its orbit, we measure the apparent position of the star again. If the star is quite close, its apparent position will have 'hopped'. From the length of the hop, it is easy to calculate how far away the star is.

Unfortunately, though, the parallax method works only for nearby stars. For distant stars, and certainly for other galaxies, our two alternating 'eyes' would need to be much further apart than 186 million miles. We have to find another method. You might think you could do it by measuring how brightly the galaxy seems to shine: surely a more distant galaxy should be dimmer than a closer one? The trouble is that the two galaxies might *really* be of different brightnesses. It's like estimating how far away a lit candle is. If some candles are brighter than others, how would you know whether you were looking at a bright candle far away, or a dim candle nearby?

Fortunately, astronomers have evidence that certain special kinds of stars are what they call 'standard candles'. They understand enough of what is going on in these stars to know how bright they are – not as we see them, but their actual brightness, the intensity of the light (or it might be X-rays or some other kind of radiation that we can measure) before it starts its long journey to our telescopes. They also know how to identify these special 'candles'; and so, as long as they can find at least one of them in a galaxy, astronomers can use it, with the assistance of well-established mathematical calculations, to estimate how far away the galaxy is.

So we have the parallax method for measuring very short distances; and there is a 'ladder' of various kinds of standard candles that we can use for measuring a range of increasingly great distances, stretching out even to very distant galaxies.

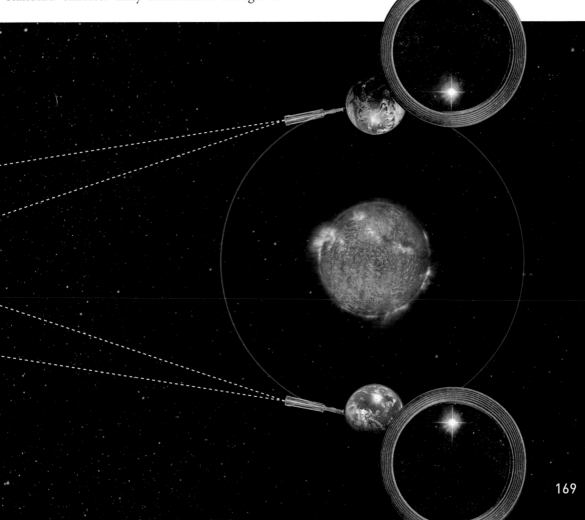

Rainbows and red shift

OK, so now we know what a galaxy is, and how to find out its distance from us. For the next step in the argument, we need to make use of the light spectrum, which we met in Chapter 7 on the rainbow. I was once asked to contribute a chapter to a book in which scientists were invited to nominate the most important invention ever. It was fun, but I had left it rather late before joining the party and all the obvious inventions had already been taken: the wheel, the printing press, the telephone, the computer and so on. So I chose an instrument that I was pretty sure nobody else would choose, and is certainly very important even though not many people have ever used one (and I must confess that I've never used one myself). I chose the *spectroscope*.

A spectroscope is a rainbow machine. If it is attached to a telescope, it takes the light from one particular star or galaxy and spreads it out as a spectrum, just as Newton did with his prism. But it is more sophisticated than Newton's prism, because it allows you to make exact measurements along the spread-out spectrum of starlight. Measurements of what? What is there to measure in a rainbow? Well, this is where it starts to get really interesting. The light from different stars produces 'rainbows'

that are different in very particular ways, and this can tell us a lot about the stars.

Does this mean that starlight has a whole variety of strange new colours, colours that we never see on Earth? No, definitely not. You have already seen, on Earth, all the colours that your eyes are capable of seeing. Do you find that disappointing? I did, when I first understood it. When I was a child, I used to love Hugh Lofting's Doctor Dolittle books. In one of the books the doctor flies to the moon, and is enchanted to behold a completely new range of colours, never before seen by human eyes. I loved this thought. For me it stood for the exciting idea that our own familiar Earth may not be typical of everything in the universe. Unfortunately, though the idea is worthwhile, the story was not true – *could not be* true. That follows from Newton's discovery that the colours we see are all contained in white light and are all revealed when white light is spread out by a prism. There are no colours outside the range we are used to. Artists

may come up with any number of different tints and shades, but all these are combinations of those basic component colours of white light. The colours we see inside our heads are really just labels made up by the brain to identify light of different wavelengths. We've already encountered the complete range of wavelengths here on Earth. Neither the moon nor the stars have any surprises to offer in the colour department. Alas.

So what did I mean when I said that different stars produce different rainbows, with differences we can measure using a spectroscope? Well, it turns out that when starlight is splayed out by a spectroscope, strange patterns of thin black lines appear in very particular places along the spectrum. Or sometimes the lines are not black but coloured, and the background is black – a difference that I'll explain in a moment. The pattern of lines looks like a barcode, the sort of barcode you see on things you buy in shops to identify them at the cash till. Different stars have the same rainbow but different patterns of lines across it – and this pattern really is a kind of barcode, because it tells us a lot about the star and what it is made of.

It isn't only starlight that shows the barcode lines. Lights on Earth do too, so we've been able to investigate, in the laboratory, what makes them. And what makes the barcodes, it turns out, is different *elements*. Sodium, for example, has prominent lines in the yellow part of the spectrum.

Sodium light (produced by an electric arc in sodium vapour) glows yellow. The reason for this is understood by physical scientists, but not by me because I'm a biological scientist who doesn't understand quantum theory.

When I went to school in the city of Salisbury in southern England, I remember being utterly fascinated by the weird sight of my bright red school cap in the yellow light of the street lamps. It didn't look red any more, but a yellowish brown. So did the bright red double-decker buses. The reason was this. Like many other English towns in those days, Salisbury used sodium vapour lamps for its street lights. These give off light only in the narrow regions of the spectrum covered by sodium's characteristic lines, and by far the brightest of sodium's lines are in the yellow. To all intents and purposes, sodium lights glow with a pure yellow light, very different from the white of sunlight or the vaguely yellowish light of an ordinary electric bulb. Since there was virtually no red at all in the light supplied by the sodium lamps, no red light could be reflected from my cap. If you are wondering what makes a cap, or a bus, red in the first place, the answer is that the molecules of dye, or paint, absorb most of the light of all colours except red. So in white light, which contains all wavelengths, mostly red light is reflected. Under sodium vapour street lamps, there is no red light to be reflected – hence the yellowy brown colour.

410 420 430 440 450 460 470 480 490 500 510 520 530 540 550 560 570 580 590 600 610 620 630 640 650 660 670 680 690

1   H   Hydrogen

Sodium is just one example. You'll remember from Chapter 4 that every element has its own unique 'atomic number', which is the number of protons in its nucleus (and also the number of electrons orbiting it). Well, for reasons connected with the orbits of its electrons, every element also has its own unique effect upon light. Unique like a barcode . . . in fact, a barcode is pretty much what the pattern of lines in the spectrum of starlight is. You can tell which of the 92 naturally occurring elements are present in a star by spreading the star's light out in a spectroscope and looking at the barcode lines in the spectrum.

There's a website where you can choose any element you like and look at its spectral barcode: **www.booksattransworld.co.uk/dawkins-elements** Just move the slider along until you come to the element you want. They are arranged in order of atomic number, from hydrogen upwards.

For example, above is the picture for hydrogen, element 1 (because it has only one proton, you'll remember). You can see that hydrogen produces four bars, one in the violet part of the spectrum, one in the dark blue, one in the pale blue and one in the red (the wavelengths of the different colours are given at the top).

In order to understand the pictures on this website, we need to be clear about a pair of otherwise confusing details. First, notice the two ways in which the bars appear: as coloured lines on a black background (in the upper part of the picture) and as black lines on a coloured background (in the lower part of the picture). These are called the emission spectrum (coloured on black background) and the absorption spectrum (black on coloured background). Which you get depends upon whether the element concerned is glowing with light (as when the element sodium glows in a sodium street lamp) or is getting in the way of light (as is often the case when an element is present in a star). I'm not going to bother with this distinction. The important point is that the bars appear in the same places along the spectrum in both cases. The barcode pattern is the same, for any particular element, whether the lines are black or coloured.

The other complicating detail is that some bars are much more prominent than others. When looking at the light from a star with a spectroscope, we usually see only the very prominent bars. But that website gives all of the lines, including the faint ones that may be seen in the lab but don't usually show up in starlight. Sodium is a good example. For practical purposes, sodium light is yellow and its prominent bars appear in the yellow part of the spectrum: you can forget the other bars, although it's interesting that they are there, as they make the patterns look even more like barcodes.

Here's the emission spectrum of sodium, with only the three strongest barcode lines shown. You can see how yellow dominates.

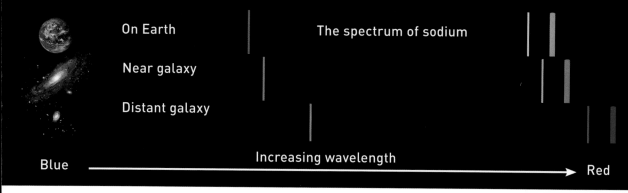

On Earth

The spectrum of sodium

Near galaxy

Distant galaxy

Blue — Increasing wavelength → Red

So, since every element has a different barcode pattern, we can look at the light from any star and see which elements are present in that star. Admittedly, it is quite tricky because the barcodes of several different elements are likely to be muddled up together. But there are ways of sorting them out. What a wonderful tool the spectroscope is!

It gets even better. The sodium spectrum at the bottom of the page opposite is what you see if you look at the light from a Salisbury street lamp, or from a star that is not very far away. Most of the stars we see – for example, the stars in the well-known constellations of the zodiac – are in our own galaxy. And the picture shown here of the spectrum of sodium light is what you see if you look at any of those. But if you look at the sodium spectrum from a star in a different galaxy, you get a fascinatingly different picture. At the top of this page is the barcode pattern of sodium light from three different places: on Earth (or from a nearby star), from a distant star in a nearby galaxy, and from a very distant galaxy.

Look first at the barcode pattern of sodium light from the distant galaxy (the bottom image), and compare it with the barcode produced by sodium light on Earth (the top image). You can see the same pattern of bars, spaced the same distance from each other. But the whole pattern is shifted towards the red end of the spectrum. How do we know it is still sodium, then? The answer is because the pattern of spacing between the bars is the same. That might not seem totally convincing if it only happened with sodium. But the same

thing happens with all the elements. In every case we see the same spacing pattern, characteristic of the element concerned, but shifted bodily along the spectrum towards the red end. What's more, for any given galaxy, all the barcodes are shifted the same distance along the spectrum.

If you look at the middle image, showing the sodium barcode in light from a galaxy that is somewhat close to ours – closer than the very distant galaxies I talked about in the previous paragraph but further away than the stars in our own Milky Way galaxy – you see an intermediate shift. You see the same spacing pattern, which is the signature of sodium, but not shifted so far. The first line is shifted along the spectrum away from deep blue, but not as far as green: only as far as light blue. And the two lines in the yellow (which combine to make the yellow colour of the Salisbury street lamps) are both shifted in the same direction, towards the red end of the spectrum, but not all the way into the red as they are in light from the distant galaxy: only a little way into the orange.

Sodium is just one example. Any other element shows the same shift along the spectrum in the red direction. The more distant the galaxy, the greater the shift towards the red. This is called the 'Hubble shift', because it was discovered by the great American astronomer Edwin Hubble, who also gave his name, after his death, to the Hubble telescope – which was used, incidentally, to photograph the very distant galaxies shown on page 167. It is also called a 'red shift', because the shift is along the spectrum in the direction of red.

Backwards to the big bang

What does the red shift mean? Fortunately, scientists understand it well. It is an example of what is called a 'Doppler shift'. Doppler shifts can happen wherever we have waves – and light, as we saw in the previous chapter, consists of waves. It's often called the 'Doppler effect' and it is more familiar to us from sound waves. When you are standing at a roadside watching the cars whizz by at high speed, the sound of every car's engine seems to drop in pitch as it passes you. You know the car's engine note really stays the same, so why does the pitch seem to drop? The answer is

the Doppler shift, and the explanation for it is as follows.

Sound travels through the air as waves of changing air pressure. When you listen to the note of a car engine – or let's say a trumpet, because it is more pleasant than an engine – sound waves travel through the air in all directions from the source of the sound. Your ear happens to lie in one of those directions, it picks up the changes in air pressure produced by the trumpet, and your brain hears them as sound. Don't imagine molecules of air flowing from the trumpet all

the way to your ear. It isn't like that at all: that would be a wind, and winds travel in one direction only, whereas sound waves travel outwards in all directions, like the waves on the surface of a pond when you drop a pebble in.

The easiest kind of wave to understand is the so-called Mexican Wave (above), in which people in a large sports stadium stand up and then sit down again in order, each person doing so immediately after the person on one side of them (say their left side). A wave of standing and then sitting moves swiftly around the stadium.

Nobody actually moves from their place, yet the wave travels. Indeed, the wave travels far faster than anybody could run.

What travels in the pond is a wave of changing height in the surface of the water. The thing that makes it a wave is that the water molecules themselves are not rushing outwards from the pebble. The water molecules are just going up and down, like the people in the stadium. Nothing really travels outwards from the pebble. It only looks like that because the high points and low points of the water move outwards.

Sound waves are a bit different. What travels in the case of sound is a wave of changing air pressure. The air molecules move a little bit, to and fro, away from the trumpet, or whatever is the source of the sound, and back again. As they do so, they knock against neighbouring air molecules and set them moving backwards and forwards too. Those in turn knock against their neighbours and the result is that a wave of molecule-knocking – which amounts to a wave of changing pressure – travels outwards from the trumpet in all directions. And it is the wave that travels from the trumpet to your ear, not the air molecules themselves. The wave travels at a fixed speed, regardless of whether the source of the sound is a trumpet or a speaking voice or a car: about 768 miles per hour in air (four times faster under water, and even faster in some solids). If you play a higher note on your trumpet, the speed at which the waves travel remains the same, but the distance between the wave crests (the *wavelength*) becomes shorter. Play a low note, and the wave crests space out more but the wave still travels at the same speed. So high notes have a shorter wavelength than low ones.

That is what sound waves are. Now for the Doppler shift. Imagine that a trumpeter standing on a snow-covered hillside plays a long, sustained note. You get on a toboggan and speed past the trumpeter (I chose a toboggan rather than a car because it is quiet, so you can hear the trumpet). What will you hear? The successive wave crests leave the trumpet at a definite distance from each other, defined by the note the trumpeter chose to play. But when you are whizzing towards the trumpeter, your ear will gobble up the successive wave crests at a higher rate than if you were standing still on the hilltop. So the trumpet's note will sound higher than it really is. Then, after you have whizzed past the trumpeter, your ear will hit the successive wave crests at a lower rate (they'll seem more spaced out, because each wave crest is travelling in the same direction as your toboggan), so the apparent pitch of the note will be lower than it really is. The same thing works if your ear is still and the source of the sound moves. It is said (I don't know whether it is true, but it is a nice story) that Christian Doppler, the Austrian scientist who discovered the effect, hired a brass band to play

on an open railway truck, in order to demon-strate it. The tune the band was playing suddenly dropped into a lower key as the train puffed past the amazed audience.

Light waves are different again – not really like a Mexican Wave and not really like sound waves. But they do have their own version of the Doppler effect. Remember that the red end of the spectrum has a longer wavelength than the blue end, with green in the middle. Suppose the bands-men on Christian Doppler's railway truck are all wearing yellow uniforms. As the train speeds to-wards you, your eyes 'gobble up' the wave crests at a faster rate than they would if the train was still. So there is a slight shift in the colour of the uniform towards the green part of the spectrum. Now, when the train goes past you and is speeding away from you, the opposite happens, and the band uniforms appear slightly redder.

There's only one thing wrong with this illus-tration. In order for you to notice the blue shift or the red shift, the train would have to be travell-ing at millions of miles per hour. Trains don't travel anywhere near fast enough for the Doppler effect on colour to be noticed. But galaxies do. The shift of the spectrum towards the red end, which you can clearly see in the positions of the sodium barcode lines in the picture on page 172, shows that very distant galaxies are travelling away from us at a rate of hundreds of millions of miles per hour. And the more distant they are (as measured by the 'standard candles' that I mentioned above), the faster they are travelling away from us (the greater the red shift).

All the galaxies in the universe are rushing away from each other, which means that they are rushing away from us too. It doesn't matter which direction you point your telescope in, the more distant galaxies are moving away from us (and from one another) at ever-increasing speed. The entire universe – space itself – is expanding at a colossal rate.

In that case, you might ask, why is it only at the level of galaxies that space is seen to expand? Why don't the stars within a galaxy rush away from each other? Why aren't you and I rushing away from each other? The answer is that clusters of things that are close to each other, like everything in a galaxy, feel the strongest pull from the gravity of their neighbours. This holds them together, while distant objects – other galaxies – recede with the expansion of the universe.

And now here is something amazing. Astronomers have looked at the expansion and worked backwards through time. It is as though they constructed a movie of the expanding universe, with the galaxies rushing apart, and then ran the film in reverse. Instead of hurtling away from each other, in the backwards film the galaxies converge. And from that film the astronomers can calculate back to the moment when the expansion of the universe must have begun. They can even calculate when that moment was. That's how they know it was somewhere between 13 and 14 billion years ago. That was the moment when the universe itself began – the moment called the

'big bang'.

Today's 'models' of the universe assume that it wasn't only the universe that began with the big bang: time itself and space itself began with the big bang too. Don't ask me to explain that, because, not being a cosmologist, I don't understand it myself. But perhaps you can now see why I nominated the spectroscope as one of the most important inventions ever. Rainbows are not just beautiful to look at. In a way, they tell us when everything began, including time and space. I think that makes the rainbow even more beautiful.

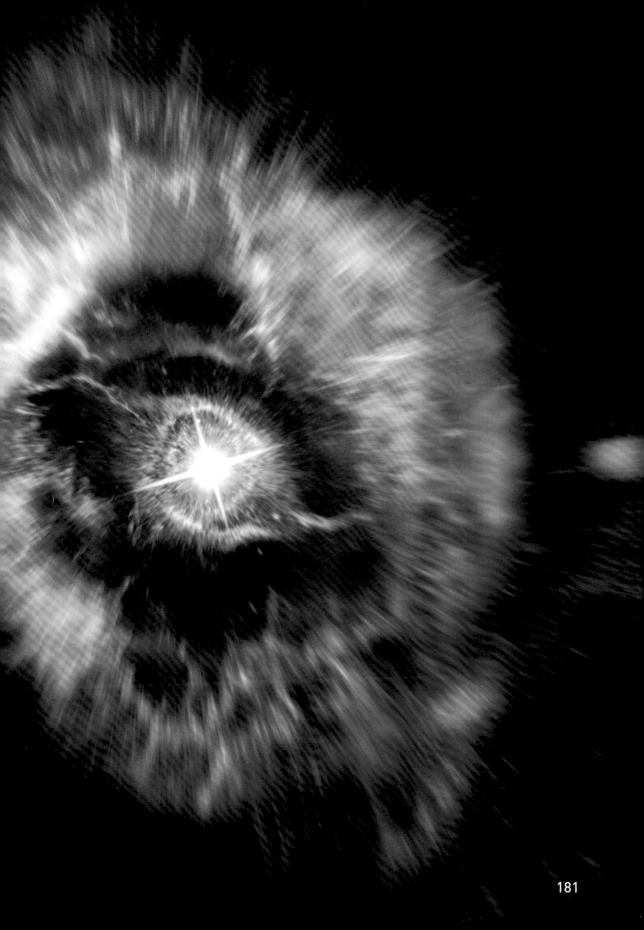

9 ARE WE

alone?

SO FAR AS I know there are few, if any, ancient myths about alien life elsewhere in the universe, perhaps because the very idea of there being a universe vastly bigger than our own world hasn't been around all that long. It took until the 1500s for scientists to see clearly that the Earth orbits the sun, and that there are other planets that do so too. But the distance and number of the stars, let alone other galaxies, was unknown and undreamed of until relatively

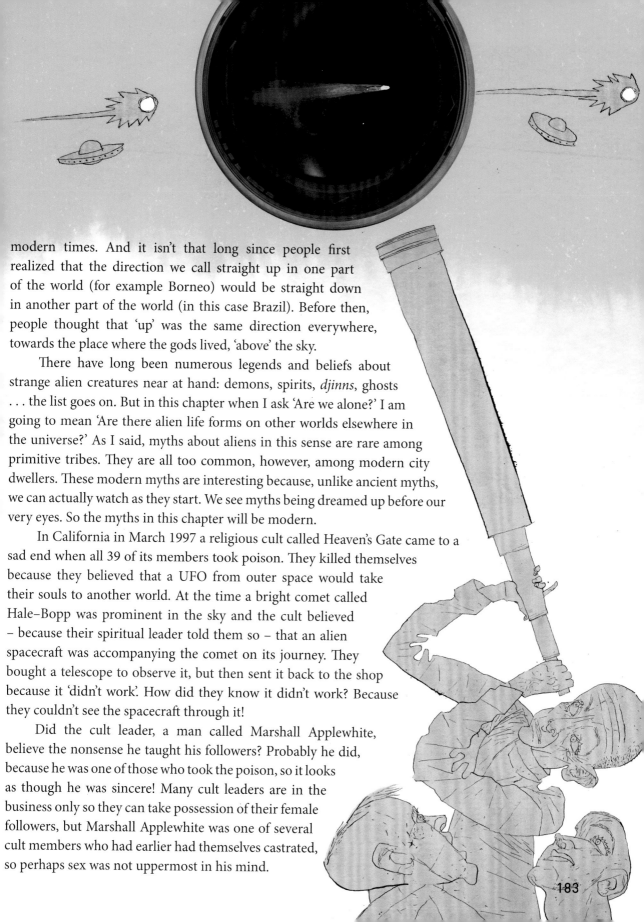

modern times. And it isn't that long since people first realized that the direction we call straight up in one part of the world (for example Borneo) would be straight down in another part of the world (in this case Brazil). Before then, people thought that 'up' was the same direction everywhere, towards the place where the gods lived, 'above' the sky.

There have long been numerous legends and beliefs about strange alien creatures near at hand: demons, spirits, *djinns*, ghosts . . . the list goes on. But in this chapter when I ask 'Are we alone?' I am going to mean 'Are there alien life forms on other worlds elsewhere in the universe?' As I said, myths about aliens in this sense are rare among primitive tribes. They are all too common, however, among modern city dwellers. These modern myths are interesting because, unlike ancient myths, we can actually watch as they start. We see myths being dreamed up before our very eyes. So the myths in this chapter will be modern.

In California in March 1997 a religious cult called Heaven's Gate came to a sad end when all 39 of its members took poison. They killed themselves because they believed that a UFO from outer space would take their souls to another world. At the time a bright comet called Hale–Bopp was prominent in the sky and the cult believed – because their spiritual leader told them so – that an alien spacecraft was accompanying the comet on its journey. They bought a telescope to observe it, but then sent it back to the shop because it 'didn't work'. How did they know it didn't work? Because they couldn't see the spacecraft through it!

Did the cult leader, a man called Marshall Applewhite, believe the nonsense he taught his followers? Probably he did, because he was one of those who took the poison, so it looks as though he was sincere! Many cult leaders are in the business only so they can take possession of their female followers, but Marshall Applewhite was one of several cult members who had earlier had themselves castrated, so perhaps sex was not uppermost in his mind.

One thing most such people seem to have in common is a love of science fiction. The members of the Heaven's Gate cult were obsessed with *Star Trek*. Of course, there is no shortage of science fiction stories about aliens from other planets, but most of us know that's just what they are: fiction, imagined, invented stories, not accounts of things that actually happened. But there are quite a lot of people who firmly, sincerely and unshakeably believe that they have personally been captured ('abducted') by aliens from outer space. So eager are they to believe this that they will do so on the flimsiest of 'evidence'. One man, for instance, believed he had been abducted, for no better reason than that he often got nosebleeds. His theory was that the aliens had put a radio transmitter in his nose to spy on him. He also thought he might be part alien himself, on the grounds that his colouring was a little darker than his parents'. A surprisingly large number of Americans, many of them otherwise normal, sincerely believe that they personally have been taken aboard flying saucers and been the victims of horrific experiments conducted by little grey men with large heads and huge, wraparound eyes. There is a whole mythology of 'alien abductions', which is as rich, as colourful and as detailed as the mythology of ancient Greece and the gods of Mount Olympus. But these alien abduction myths are recent, and you can actually go and talk to people who believe they have been abducted: apparently normal, sane, level-headed people, who will tell you they saw the aliens face to face; actually tell you what the aliens look like, and what they say while performing their nasty experiments and sticking needles into people (the aliens speak English, of course!).

Susan Clancy is one of several psychologists who have made detailed studies of people who claim to have been abducted. Not all of them have clear memories of the event, or even any memories at all. They account for this by saying that obviously the aliens must have used some devilish technique to wipe their memories clean after they had finished experimenting on their bodies. Sometimes they go to a hypnotist, or a psychotherapist of some kind, who helps them to 'recover their lost memories'.

Recovering 'lost' memory is a whole other story, by the way, which is interesting in its own right. When we think we remember a real incident, we may only be remembering another memory . . . and so on back to what may or may not have been a real incident originally. Memories of memories of memories can become progressively distorted. There is good evidence that some of our most vivid memories are actually *false* memories. And false memories can be deliberately planted by unscrupulous 'therapists'.

False memory syndrome helps us understand why at least some of the people who think they have been abducted by aliens claim to have such vivid memories of the incident. What usually happens is that a person becomes obsessed with aliens through reading stories in the newspapers about other alleged abductions. Often, as I said, these people are fans of *Star Trek*, or other science fiction tales. It is a striking fact that the aliens they think they've met usually look very like the ones portrayed in the most recent television fiction about aliens, and they usually do the same kind of 'experiments' as have recently been seen on television.

The next thing that may happen is that the person is afflicted by a frightening experience called sleep paralysis. It is not uncommon. You may even have experienced it yourself, in which case I hope it will be a bit less scary the next time it happens if I explain it to you now. Normally, when you are asleep and dreaming, your body is paralysed. I suppose it's to stop your muscles working in tune with your dreams and making you sleepwalk (though this does, of course, sometimes happen). And normally, when you wake and your dream vanishes, the paralysis goes and you can move your muscles.

185

But occasionally there is a delay between your mind returning to consciousness and your muscles coming back to life, and that is called sleep paralysis. It is frightening, as you can imagine. You are sort of awake, and you can see your bedroom and everything in it, but you can't move. Sleep paralysis is often accompanied by terrifying hallucinations. People feel surrounded by a sense of dreadful danger, which they can't put a name to. Sometimes they even see things that are not there, just as in a dream. And, also as in a dream, to the dreamer they seem absolutely real.

Now, if you are going to have a hallucination when you suffer sleep paralysis, what might that hallucination look like? A modern science fiction fan might well see little grey men with big heads and wraparound eyes. In earlier centuries, before science fiction came along, the visions people saw were different: hobgoblins, perhaps, or werewolves; bloodsucking vampires or (if they were lucky) beautiful winged angels.

The point is that the images people see when experiencing sleep paralysis are not really there but are conjured up in the mind from past fears, legends or fiction. Even if they don't hallucinate, the experience is so frightening that, when they finally wake up, sleep paralysis victims often believe that something horrible has happened to them. If you are primed to believe in vampires, you might wake with a strong belief that a bloodsucker has attacked you. If I am primed to believe in alien abductions I might wake up believing that I was abducted and my memory then wiped clean by aliens.

The next thing that typically happens to sleep paralysis victims is that, even if they didn't actually hallucinate aliens and gruesome experiments at the time, their fearful reconstruction of what they suspect may have happened becomes consolidated as a false memory. This process is often helped along by friends and family, who eagerly pump them for more and more detailed accounts of what happened, and even prompt them with leading questions: 'Were there aliens there? What colour were they? Were they grey? Did they have big wraparound eyes like in the movies?' Even questions can be enough to implant or cement a false memory. When you look at it like this, it is not so surprising that a 1992 poll concluded that nearly four million Americans thought they had been abducted by aliens.

My friend the psychologist Sue Blackmore points out that sleep paralysis was the most likely cause of earlier imagined horrors, too, before the idea of space aliens became popular. In medieval times people claimed to have been visited in the middle of the night by an 'incubus' (a male demon visiting a female victim to have sex with her) or 'succubus' (a female demon visiting a male victim to have sex with him). One of the effects of sleep paralysis is that, if you try to move, it feels as

186

though something is pressing down on your body. This could easily be interpreted by the terrified victim as a sexual assault. Legend in Newfoundland talks of an 'Old Hag' who visits people in the night and presses down on their chests. And there is a legend in Indochina of a 'Grey Ghost' who visits people in the dark and paralyses them.

So we have a good understanding of why people believe they have been abducted by aliens, and we can tie the modern myths of alien abduction in with earlier myths of rapacious incubi and succubi, or of vampires with long canine teeth who visit in the night and suck our blood. There is no good evidence at all that this planet has ever been visited by aliens from outer space (or, for that matter, by incubi or succubi or demons of any kind). But we are still left with the question of whether there actually are living things on other planets. Just because they haven't visited us it doesn't mean they don't exist. Could the same process of evolution, or even a very different process that perhaps resembles our kind of evolution only slightly, have got going on other planets as well as ours?

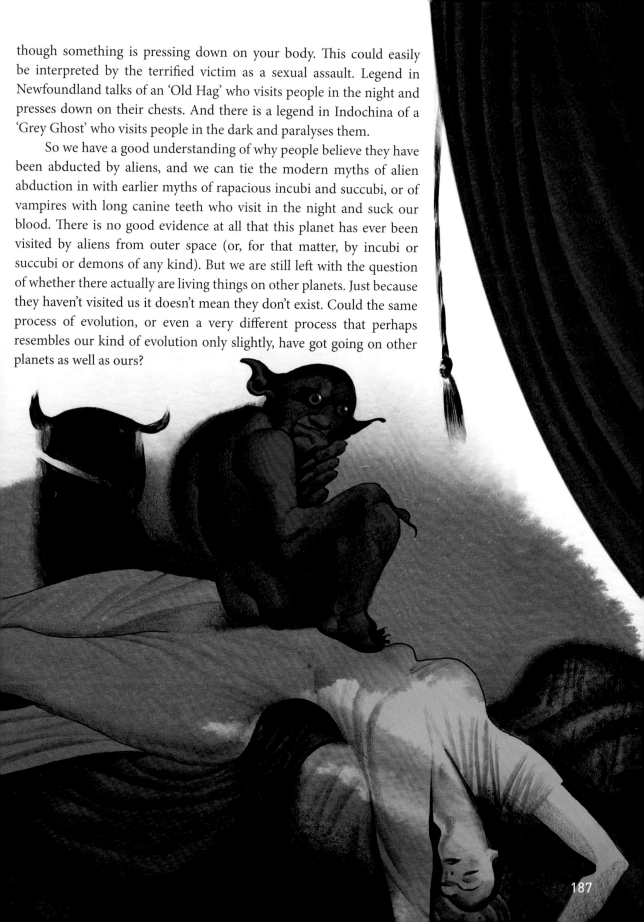

IS THERE **REALLY** LIFE ON OTHER PLANETS?

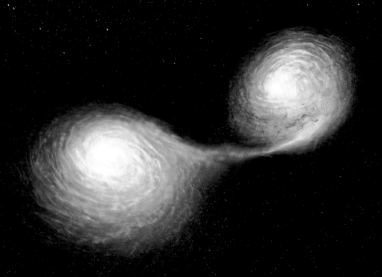

NOBODY KNOWS. If you forced me to give an opinion one way or the other, I'd say yes, and probably on millions of planets. But who cares about an opinion? There is no direct evidence. One of the great virtues of science is that scientists know when they don't know the answer to something. They cheerfully admit that they don't know. Cheerfully, because not knowing the answer is an exciting challenge to try to find it.

One day we may have definite evidence of life on other planets, and then we'll know for sure. For now, the best a scientist can do is write down the kind of information that might reduce the uncertainty, might take us from guesswork to an estimate of likelihood. And that, in itself, is an interesting and challenging thing to do.

The first thing we might ask is how many planets there are. Until quite recently, it was possible to believe that the ones orbiting our sun were the only ones, because planets could not be detected by even the largest telescopes. Nowadays we have good evidence that lots of stars have planets, and new 'extra-solar' planets are discovered almost every day. An extra-solar planet is a planet orbiting a star other than the sun (*sol* is the Latin for sun and *extra* is the Latin for outside).

You might think that the obvious way to detect a planet is to see it through a telescope. Unfortunately, planets are too dim to be seen at any great distance – they don't glow in their own right but only reflect their star's light – so we can't see them directly. We have to rely on indirect methods, and the best method again makes use of the spectroscope, the instrument we met in Chapter 8. Here's how.

When a heavenly body orbits another one of approximately equal size, they orbit each other, because they exert approximately equal gravitational force on each other. Several of the bright stars that we see when we look up are actually two stars – so-called binaries – in orbit around each other like the two ends of a dumbbell connected by an invisible rod. When one body is much smaller than the other, as is the case with a planet and its star, the smaller one whizzes around the larger one, while the larger one makes only little token movements in response to the gravitational pull of the smaller. We say that Earth orbits the sun, but actually the sun also makes tiny movements in response to the gravity of Earth.

And a planet as large as Jupiter can have an appreciable effect on the position of its star. These token movements of a star are too small to count as 'going round' the planet, but they are large enough to be detected by our instruments, even though we can't see the planet at all.

How we detect these movements is interesting in its own right. Any star is too far away for us to be able to see it actually moving, even with a powerful telescope. But, strangely, although we can't see a star move, we can measure the speed with which it does so. That sounds odd, but this is where the spectroscope comes in. Remember the Doppler shift from Chapter 8? When the star's movement happens to be away from us the light from it will be red-shifted. When the star's movement is towards us its light will be blue-shifted. So, if a star has an orbiting planet the spectroscope will show us a rhythmically pulsating red-blue-red-blue shift pattern, and the timing of these regular shifts will tell us the length of the planet's year. Of course it's complicated when there's more than one planet. But astronomers are good at mathematics and they can cope with that complication. At the time of writing (January 2011) 484 planets have been detected by this means, orbiting 408 stars. There will surely be more by the time you read this.

There are other methods of detecting planets. For example, when a planet passes across the face of its star, a small portion of the face of the star is obscured or eclipsed – like when we see the moon eclipsing the sun, except that the moon looks much bigger because it is so much closer

10,000

When a planet comes between us and its star, the star becomes very very slightly dimmer, and sometimes our instruments are sensitive enough to detect this dimming. So far, 110 planets have been discovered in this way. And there are a few other methods, too, which have detected another 35 planets. Some planets have been detected by more than one of these techniques, and the present grand total is 519 planets orbiting stars in our galaxy other than the sun.

In our galaxy, the great majority of stars where we have looked for planets have turned out to possess them. So, assuming our galaxy is typical, we can probably conclude that most of the stars in the universe have planets in orbit around them. The number of stars in our galaxy is about 100 billion, and the number of galaxies in the universe is about the same again. That means something like 10,000 billion billion stars in total. About 10 per cent of known stars are described by astronomers as 'sun-like'. Stars that are very different from the sun, even if they have planets, are unlikely to support life on those planets for various reasons: for example,

stars that are much bigger than the sun tend not to last long enough before exploding. But even if we confine ourselves to the planets orbiting sun-like stars we are likely to be dealing in billions of billions – and that would probably still be an underestimate.

All right, but how many of those planets orbiting the 'right kind of star' are likely to be suitable for supporting life? The majority of extra-solar planets discovered so far are 'Jupiters'. That means they are 'gas giants', mostly made of gas at high pressure. This is not surprising, as our methods of detecting planets are usually not sensitive enough to notice anything smaller than Jupiters. And Jupiters – gas giants – are not suitable for life as we know it. Of course, that doesn't mean that life as we know it is the only possible kind of life. There might even be life on Jupiter itself, although I doubt it. We don't know what proportion of those billions of billions of planets are Earth-like rocky planets, as opposed to Jupiter-like gas giants. But even if the proportion is quite low, the absolute number will still be high because the total is so huge.

000,000,
000,000,
000,000,

Looking for Goldilocks

Life as we know it depends on water. Once again, we should beware of fixing our attention on life as we know it, but for the moment exobiologists (scientists searching for extraterrestrial life) regard water as essential – so much so that a good part of their effort is given over to searching the heavens for signs of it. Water is a lot easier to detect than life itself. If we find water it certainly doesn't mean there has to be life, but it is a step in the right direction.

For life as we know it to exist, at least some of the water has to be in liquid form. Ice won't do, nor will steam. Close inspection of Mars shows evidence of liquid water, in the past if not today. And several other planets have at least some water, even if it is not in liquid form. Europa, one of the moons of Jupiter, is covered with ice, and it has been plausibly suggested that under the ice is a sea of liquid water. People once thought Mars was the best candidate for extraterrestrial life within the solar system, and a famous astronomer called Percival Lowell even drew what he claimed were canals criss-crossing its surface. Spacecraft have now taken detailed photographs of Mars, and have even landed on its surface, and the canals have turned out to be figments of Lowell's imagination. Nowadays Europa has taken the place of Mars as the prime site of speculation about extraterrestrial life in our own solar system, but most scientists think we have to look further afield. Evidence suggests that water is not particularly rare on extra-solar planets.

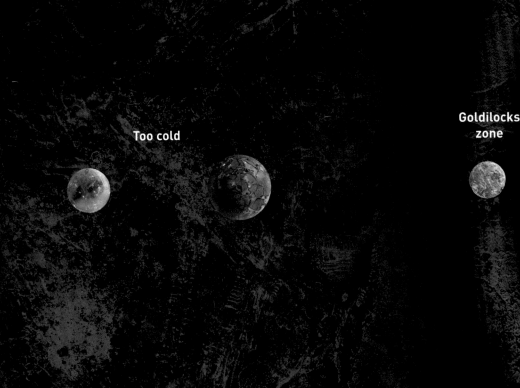

Too cold

Goldilocks zone

What about temperature? How finely tuned does the temperature of a planet have to be, if it is to support life? Scientists talk of a so-called 'Goldilocks Zone': 'just right' (like baby bear's porridge) between two wrong extremes of too hot (like father bear's porridge) and too cold (like mother bear's porridge). The orbit of Earth is 'just right' for life: not too close to the sun, where water would boil, and not too far from the sun, where all the water would freeze solid and there wouldn't be enough sunlight to feed the plants. Although there are billions and billions of planets out there, we cannot expect more than a minority of them to be just right, where temperature and distance from their star are concerned.

Recently (May 2011) a 'Goldilocks planet' was discovered orbiting a star called Gliese 581, which is about 20 light years away from us (not very far as stars go, but still a vast distance by human standards). The star is a 'red dwarf', much smaller than the sun, and its Goldilocks zone is correspondingly closer in. It has at least six planets, named Gliese 581e, b, c, g, d and f. Several of them are small, rocky planets like Earth, and one of them, Gliese 581d, is thought to be in the Goldilocks zone for liquid water. It is not known whether Gliese 581d actually has water, but if so it is likely to be liquid rather than ice or vapour. Nobody is suggesting that Gliese 581d actually does have life, but the fact that it has been discovered so soon after we started looking makes one think there are probably lots of Goldilocks planets out there.

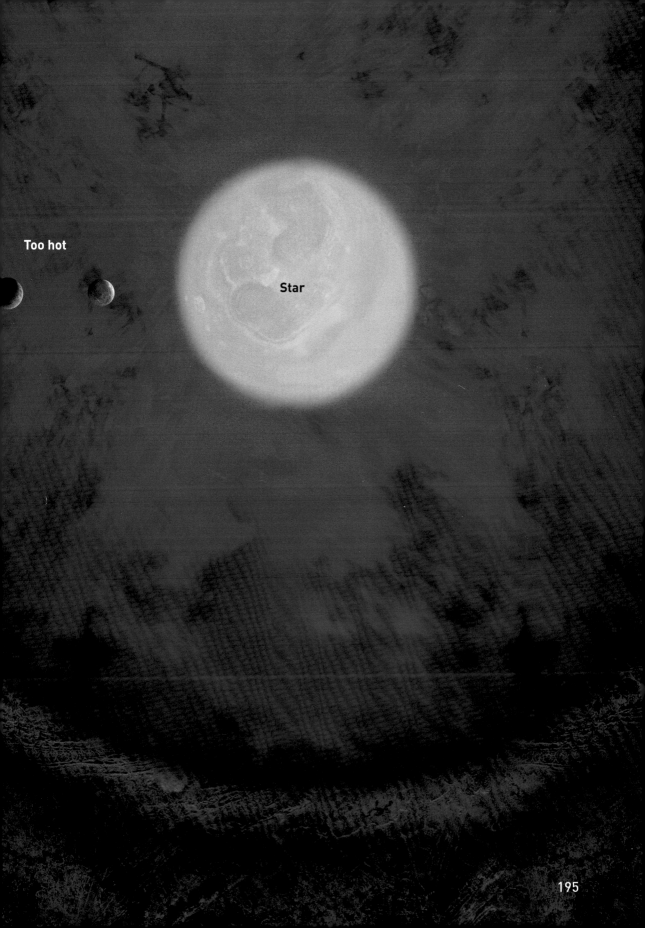

What about the size of a planet? Is there a Goldilocks size – not too big and not too small, but just right? The size of a planet – more strictly its mass – has a big impact upon life because of gravity. A planet with the same diameter as Earth, but mostly made of solid gold, would have a mass more than three times as great. The gravitational pull of the planet would be over three times as strong as we are used to on Earth. Everything would weigh more than three times as much, and that includes any living bodies on the planet. Putting one foot in front of the other would be a great labour. An animal the size of a mouse would need to have thick bones to support its body, and it would lumber about like a miniature rhinoceros, while an animal the size of a rhinoceros might suffocate under its own weight.

Just as gold is heavier than the iron, nickel and other things that Earth is mostly made of, coal is much lighter. A planet the size of Earth but mostly made of coal would have a gravitational pull only about a fifth as strong as we are used to. An animal the size of a rhinoceros could skitter about on thin, spindly legs like a spider. And animals far bigger than the largest dinosaurs could happily evolve, if the other conditions on the planet were right. The moon's gravity is about one-sixth that of Earth. That is why astronauts on the moon moved with a curious bounding gait, which looked quite comical because of their large bulk in their space suits. An animal that actually evolved on a planet with such weak gravity would be built very differently – natural selection would see to that.

If the gravitational pull were too strong, as it would be on a neutron star, there could be no life at all. A neutron star is a kind of collapsed star. As we learned in Chapter 4, matter normally consists almost entirely of empty space. The distance between atomic nuclei is vast, compared with the size of the nuclei themselves. But in a neutron star the 'collapsing' means that all that empty space has gone. A neutron star can have as much mass as the sun yet be only the size of a city, so its gravitational pull is shatteringly strong. If you were plonked down on a neutron star, you would weigh a hundred billion times what you weigh on Earth. You'd be flattened. You couldn't move. A planet would only need to have a tiny fraction of the gravitational pull of a neutron star to put it outside the Goldilocks zone – not just for life as we know it, but for life as we could possibly imagine it.

197

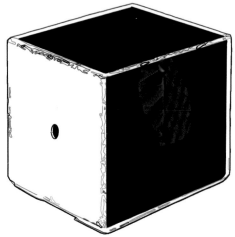

Here's looking at you

If there are living creatures on other planets, what might they look like? There's a widespread feeling that it's a bit lazy for science fiction authors to make them look like humans, with just a few things changed – bigger heads or extra eyes, or maybe wings. Even when they are not humanoid, most fictional aliens are pretty clearly just modified versions of familiar creatures, such as spiders, octopuses or mushrooms. But perhaps it is not just lazy, not just a lack of imagination. Perhaps there really is good reason to suppose that aliens, if there are any (and I think there probably are), might not look too unfamiliar to us. Fictional aliens are proverbially described as bug-eyed monsters, so I'll take eyes as my example. I could have taken legs or wings or ears (or even wondered why animals don't have wheels!). But I'll stick to eyes and try to show that it isn't really lazy to think that aliens, if there are any, might very well have eyes.

Eyes are pretty good things to have, and that is going to be true on most planets. Light travels, for practical purposes, in straight lines. Wherever light is available, such as in the vicinity of a star, it is technically easy to use light rays to find your way around, to navigate, to locate objects. Any

planet that has life is pretty much bound to be in the vicinity of a star, because a star is the obvious source of the energy that all life needs. So the chances are good that light will be available wherever life is present; and where light is present it is very likely that eyes will evolve because they are so useful. It is no surprise that eyes have evolved on our planet dozens of times independently.

There are only so many ways to make an eye, and I think every one of them has evolved somewhere in our animal kingdom. There's the camera eye (above left), which, like the camera itself, is a darkened chamber with a small hole at the front letting in light, through a lens, which focuses an upside-down image on a screen – the 'retina' – at the back. Even a lens is not essential. A simple hole will do the job if it is small enough, but that means that very little light gets through, so the image is very dim – unless the planet happens to get a lot more light from its star than we get from the sun. This is of course possible, in which case the aliens could indeed have pinhole eyes. Human eyes (opposite, right) have a lens, to increase the amount of light that is focused on the retina. The retina at the back is carpeted with cells that are sensitive to light and tell the brain about it via nerves. All vertebrates have this kind of eye, and the camera eye has been independently evolved by lots of other kinds of animals, including octopuses. And invented by human designers too, of course.

Jumping spiders (left, below) have a weird kind of scanning eye. It is sort of like a camera eye except that the retina, instead of being a broad carpet of light-sensitive cells, is a narrow strip. The strip retina is attached to muscles which move it about so that it 'scans' the scene in front of the spider. Interestingly, that is a bit like what a television camera does too, since it has only a single channel to send a whole image along. It scans across and down in lines, but does it so fast that the picture we receive looks like a single image. Jumping spider eyes don't scan so fast, and they tend to concentrate on 'interesting' parts of the scene such as flies, but the principle is the same.

Then there's the compound eye (right, below), which is found in insects, shrimps and various other animal groups. A compound eye consists of hundreds of tubes, radiating out from the centre of a hemisphere, each tube looking in a slightly different direction. Each tube is capped by a little lens, so you could think of it as a miniature eye. But the lens doesn't form a usable image: it just concentrates the light in the tube. Since each tube accepts light from a different direction, the brain can combine the information from them all to reconstruct an image: rather a crude image, but good enough to let dragonflies, for instance, catch moving prey on the wing.

Our largest telescopes use a curved mirror rather than a lens, and this principle too is used in animal eyes, specifically in scallops. The scallop eye uses a curved mirror to focus an image on a retina, which is in front of the mirror. This inevitably gets in the way of some of the light, as the equivalent does in reflecting telescopes, but it doesn't matter too much as most of the light gets through to the mirror.

That list pretty much exhausts the ways of making an eye that scientists can imagine, and all of them have evolved in animals on this planet, most of them more than once. It is a good bet that, if there are creatures on other planets that can see, they will be using eyes of a kind that we would find familiar.

Let's exercise our imaginations a bit more. On the planet of our hypothetical aliens, the radiant energy from their star will probably range from radio waves at the long end to X-rays at the short. Why should the aliens limit themselves to the narrow band of frequencies that we call 'light'? Maybe they have radio eyes? Or X-ray eyes?

A good image relies on high *resolution*. What does that mean? The higher the resolution, the closer two points can be to each other while still being distinguished from each other. Not surprisingly, long wavelengths don't make for good resolution. Light wavelengths are measured in minute fractions of a millimetre and give excellent resolution, but radio wavelengths are measured in metres. So radio waves would be lousy for forming images, although they are very good for communication purposes because they can be *modulated*. Modulated means changed, extremely rapidly, in a

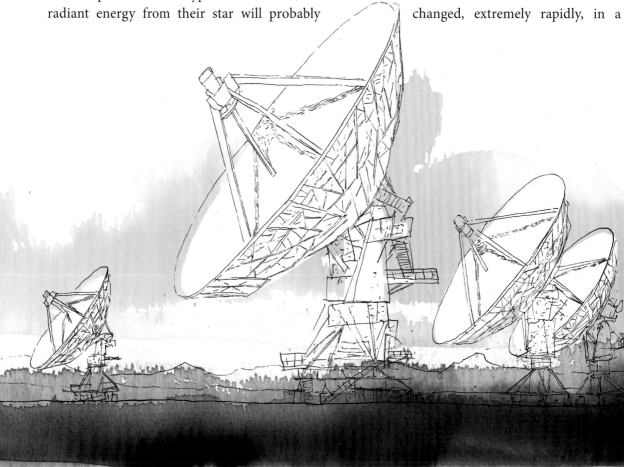

controlled way. So far as is known, no living creature on our planet has evolved a natural system for transmitting, modulating or receiving radio waves: that had to wait for human technology. But perhaps there are aliens on other planets that have evolved radio communication naturally.

What about waves shorter than light waves – X-rays, for example? X-rays are difficult to focus, which is why our X-ray machines form shadows rather than true images, but it is not impossible that some life forms on other planets have X-ray vision.

Vision of any kind depends on rays travelling in straight, or at least predictable, lines. It is no good if they are scattered every which way, as light rays are in fog. A planet that is permanently shrouded in thick fog would not encourage the evolution of eyes. Instead, it might foster the use of some kind of echo ranging system like the 'sonar' used by bats, dolphins and man-made submarines. River dolphins are extremely good at using sonar, because their water is full of dirt, which is the watery equivalent of fog. Sonar has evolved at least four times in animals on our planet (in bats, whales, and two separate kinds of cave-dwelling birds). It would not be surprising to find sonar evolving on an alien planet, especially one that is permanently shrouded in fog.

Or, if the aliens have evolved organs that can handle radio waves for communication, they might also evolve true radar to find their way around, and radar does work in fog. On our planet, there are fish that have evolved the ability to find their way about using distortions in an electric field that they themselves create. In fact, this trick has evolved twice independently, in a group of African fish and in a completely separate group of South American fish. Duck-billed platypuses have electric sensors in their bills which pick up the electrical disturbances in water caused by the muscular activity of their prey. It is easy to imagine an alien life form that has evolved electrical sensitivity along the same lines as the fishes and the platypus, but to a more advanced level.

This chapter is rather different from the others in this book because it emphasizes what we don't know, rather than what we do. Yet even though we have not yet discovered life on other planets (and indeed, may never do so), I hope you have seen and been inspired by how much science can tell us about the universe. Our search for life elsewhere is not haphazard or random: our knowledge of physics and chemistry and biology equips us to seek out meaningful information about stars and planets vast distances away, and to identify planets that are at least possible candidates as hosts for life. There is much that remains deeply mysterious, and it is not likely that we will ever uncover all the secrets of a universe as vast as ours: but, armed with science, we can at least ask sensible, meaningful questions about it and recognize credible answers when we find them. We don't have to invent wildly implausible stories: we have the joy and excitement of real scientific investigation and discovery to keep our imaginations in line. And in the end that is more exciting than fantasy.

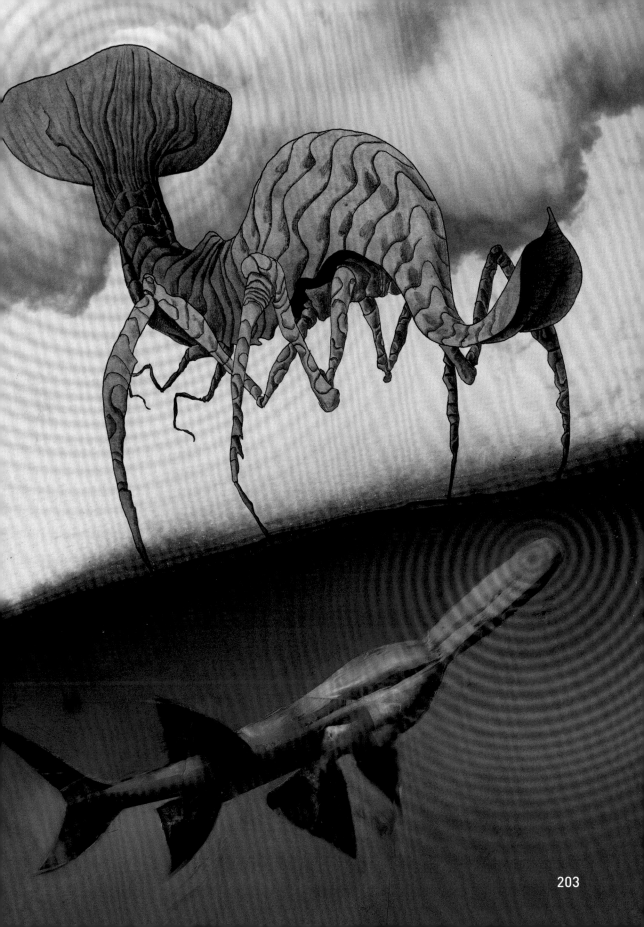

203

WHAT IS AN EARTH

IMAGINE that you are sitting quietly in your room, perhaps reading a book or watching television or playing a computer game. Suddenly there is a terrifying rumbling sound, and the whole room starts to shake. The light swings wildly from the ceiling, ornaments clatter off the shelves, furniture is hurled across the floor, you are tipped out of your chair. After two minutes or so everything settles down again and there is a blessed silence, broken only by the crying of a frightened child and the barking of a dog. You pick yourself up and think how lucky you are that the whole house didn't collapse. In a very severe earthquake, it might well have done.

While I was beginning to write this book, the Caribbean island of Haiti was hit by a devastating earthquake and the capital city, Port au Prince, was largely destroyed. Two hundred and thirty thousand people are believed to have been killed, and many others, including poor orphaned children, are still wandering the streets, homeless, or living in temporary camps.

Later, as I was revising the book, another earthquake, even stronger, occurred under the sea off the north-eastern coast of Japan. It caused a gigantic wave – a 'tsunami' – that wrought unimaginable destruction when it swept ashore, carrying whole towns with it, killing thousands of people and leaving millions homeless, and setting off dangerous explosions in a

QUAKE?

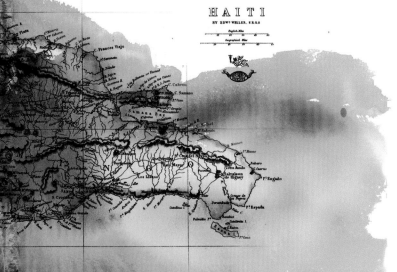

HAITI
BY EDW? WELLER, F.R.G.S.

nuclear power plant already damaged by the earthquake.

Earthquakes, and the tsunamis they cause, are common in Japan (the very word 'tsunami' was originally Japanese), but the country had experienced nothing like this in living memory. The prime minister described it as the country's worst experience since the Second World War, when atomic bombs destroyed the Japanese cities of Hiroshima and Nagasaki. Indeed, earthquakes are common all the way around the rim of the Pacific Ocean – the New Zealand city of Christchurch suffered severe damage and loss of life in a quake just one month before that which struck Japan. This so-called 'ring of fire' includes much of California and the western United States, where there was a famous earthquake in the city of San Francisco in 1906. The larger city of Los Angeles is also vulnerable.

What happens when an earthquake strikes?

YOU CAN GET some idea of what a major earthquake near Los Angeles would be like by looking at a computer simulation. This simulation is a kind of visual forecast of something that hasn't happened, but might, based on realistic science – a sort of 'as-if' film generated by the computer. The film shows you an event that hasn't actually happened, so that you can see what it would look like if it did happen – as, one day, this one probably will.

The pictures here show two sequences of still shots from the simulation. The narrow left-hand strip on each page shows the area from above, looking south to north with Los Angeles marked, like a map. The red and yellow splodge beginning near the bottom of the first two frames is where the earthquake starts. It is called the 'epicentre' of the earthquake. The thin red line snaking up the map is the San Andreas Fault, which I'll come on to in a minute. For the moment, just think of it as a gash in the ground, a line of weakness in the Earth's surface.

The wider right-hand sequence is not a map, but a view of the landscape as if seen from a plane, looking in the opposite direction towards the south-east from Los Angeles, towards the mountains and the epicentre of the earthquake (again marked in red).

If you were to run the simulation on your computer, you'd see something rather terrifying. On the map you'd see the red centre of the earthquake rushing north up the San Andreas Fault, with waves of blue, green and yellow, representing quaking of varying strengths, fanning outwards on both sides. After about 80 seconds, the red centre reaches a point opposite Los Angeles, and yellow and green waves are already passing through the city. Another 10 seconds, and red waves have reached the centre of Los Angeles. At this point you can look at the right-hand picture, the 'view from the plane', to see what is actually going on down there – and it's an extraordinary sight. The whole landscape is behaving like a liquid. It looks like the sea, with waves passing through it. Solid, dry land, with waves sweeping through it as they do on the sea! That's an earthquake.

If you were down on the ground, you wouldn't see the waves because you'd be too close to them, and too small compared with them. You'd just

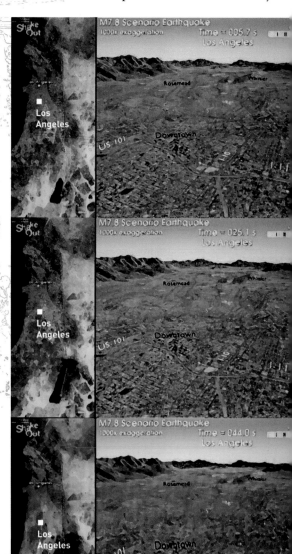

feel the ground moving and shaking beneath your feet, as I described in the opening scene of this chapter. If the shaking got really strong your house might fall down.

The colours on the simulation are called 'false colours', and they are used by the computer simply as a way of telling us how strong the quake is in different places. Blue means weak quaking, red means strong quaking, with green and yellow in between. The colours help us to visualize waves of movement through the Earth's surface – and to see how fast they travel. The 'red' centre of the earthquake roars up the San Andreas Fault at about 8,000 kilometres (5,000 miles) per hour.

As I said, this is only a computer simulation, not a real film of an earthquake. The computer has exaggerated the amount of movement, so that it looks a thousand times worse than it would be in real life. But it would still be pretty terrifying.

In a moment I'm going to explain what an earthquake really is, and what a 'fault line' is – like the San Andreas Fault, and similar ones in other parts of the world. But first, let's look at some myths.

If you have access to the Internet, look at the film here:
*www.booksattransworld.co.uk/
dawkins-earthquake*

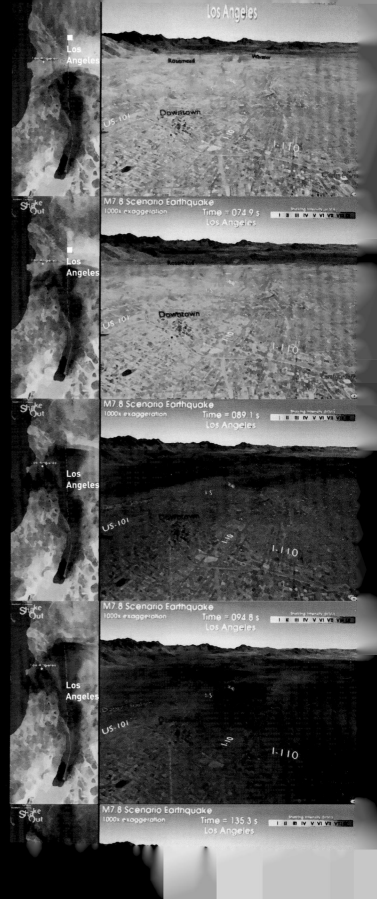

Earthquake myths

We'll begin with a pair of myths that may have grown up around particular earthquakes, earthquakes that actually happened at certain moments in history.

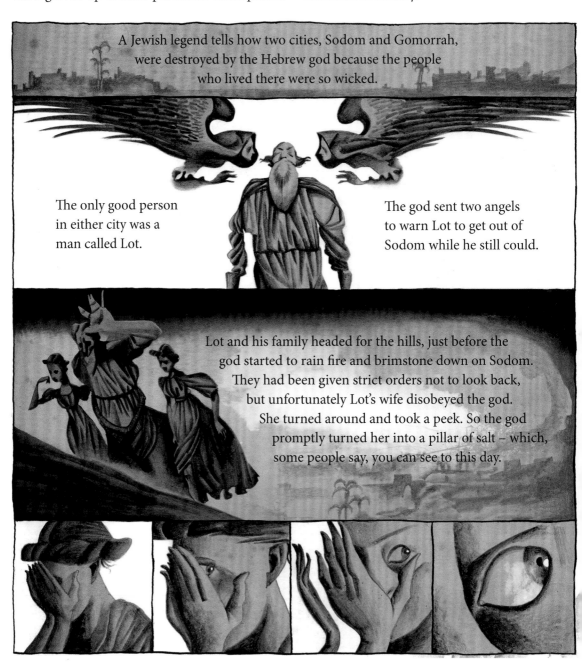

A Jewish legend tells how two cities, Sodom and Gomorrah, were destroyed by the Hebrew god because the people who lived there were so wicked.

The only good person in either city was a man called Lot.

The god sent two angels to warn Lot to get out of Sodom while he still could.

Lot and his family headed for the hills, just before the god started to rain fire and brimstone down on Sodom. They had been given strict orders not to look back, but unfortunately Lot's wife disobeyed the god. She turned around and took a peek. So the god promptly turned her into a pillar of salt – which, some people say, you can see to this day.

Some archaeologists claim to have found evidence that a large earthquake shattered the region where Sodom and Gomorrah are believed to have stood about 4,000 years ago.

If this is true, the legend of their destruction might belong in our list of earthquake myths.

Another biblical myth which might have started with a particular earthquake is the story of how Jericho was brought down. Jericho, which lies a little north of the Dead Sea in Israel, is one of the oldest cities in the world. It has suffered from earthquakes right up to recent times: in 1927 it was close to the centre of a severe one which shook the whole region and killed hundreds of people in Jerusalem, some 25 kilometres (15 miles) away.

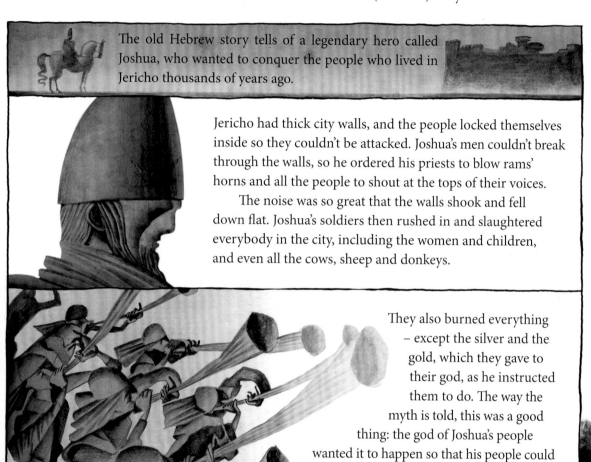

The old Hebrew story tells of a legendary hero called Joshua, who wanted to conquer the people who lived in Jericho thousands of years ago.

Jericho had thick city walls, and the people locked themselves inside so they couldn't be attacked. Joshua's men couldn't break through the walls, so he ordered his priests to blow rams' horns and all the people to shout at the tops of their voices.

The noise was so great that the walls shook and fell down flat. Joshua's soldiers then rushed in and slaughtered everybody in the city, including the women and children, and even all the cows, sheep and donkeys.

They also burned everything – except the silver and the gold, which they gave to their god, as he instructed them to do. The way the myth is told, this was a good thing: the god of Joshua's people wanted it to happen so that his people could take over all the land that had previously belonged to the people of Jericho.

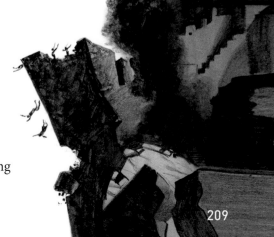

Since Jericho is such an earthquake-prone place, people nowadays have suggested that the legend of Joshua and Jericho may have begun with an ancient earthquake, which shook the city so violently that the walls fell down. You can easily imagine how a distant folk memory of a disastrous earthquake could be exaggerated and distorted as it was passed by word of mouth down through generations of people who couldn't read or write, until eventually it grew into the legend of the great tribal hero Joshua, and all that noisy shouting and horn-blowing.

The two myths I've just described may have begun with particular earthquakes in history. There are also lots of other myths, from all around the world, that have come into being as people have tried to understand what earthquakes are in general.

Since Japan experiences so many earthquakes, it's not surprising that Japan has some pretty colourful earthquake myths.

According to one of these, the land floated on the back of a gigantic catfish called Namazu. Whenever Namazu flipped his tail, the Earth would shake.

Many thousands of miles south, the Maoris of New Zealand, who arrived by canoe and settled there a few centuries before European sailors arrived, believed that Mother Earth was pregnant with her child, the god Ru. Whenever baby Ru kicked or stretched inside his mother's womb, there was an earthquake.

Back in the north, some Siberian tribes believed that the Earth sat on a sledge, pulled by dogs and driven by a god called Tull. The poor dogs had fleas, and when they scratched there was an earthquake.

In one West African legend, the Earth is a disc, held up on one side by a great mountain and on the other side by a monstrous giant, whose wife holds up the sky. Every so often the giant and his wife hug each other, and then, as you can well imagine, the Earth moves.

Other West African tribes believed that they lived on top of a giant's head. The forest was his hair, and the people and animals were like fleas wandering around on his head.

Earthquakes were what happened when the giant sneezed. At least, that is what they were supposed to believe, though I rather doubt they really did.

Nowadays we know what earthquakes really are, and it is time to put away the myths and look at the truth.

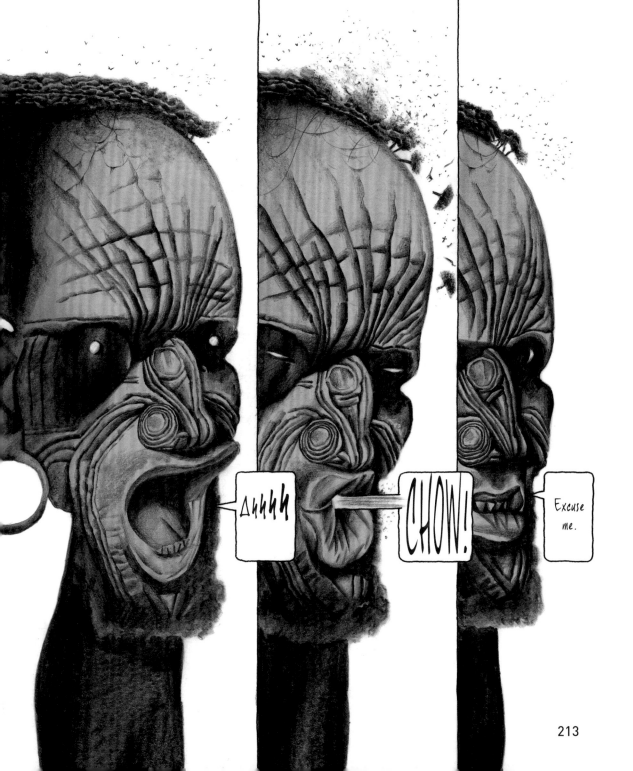

213

WHAT EARTHQUAKES REALLY ARE!

First, we need to hear the remarkable story of plate tectonics.

Everybody knows what a map of the world looks like. We know the shape of Africa and the shape of South America, and we know that the wide Atlantic Ocean separates them. We can all recognize Australia, and we know that New Zealand lies to the south-east of Australia. We know that Italy looks like a boot, about to kick the 'football' of Sicily, and some people think New Guinea looks like a bird. We can easily recognize the outline of Europe, even though the borders within it change all the time. Empires come and go; the frontiers between countries are shifted again and again through history. But the outlines of the continents themselves stay fixed. Don't they? Well, no, they don't, and that is the big point. They move, although admittedly very slowly, and so do the positions of the mountain ranges: the Alps, the Himalayas, the Andes, the Rockies. To be sure, these great geographical features are fixed on the timescale of human history. But the Earth

The world today ▼

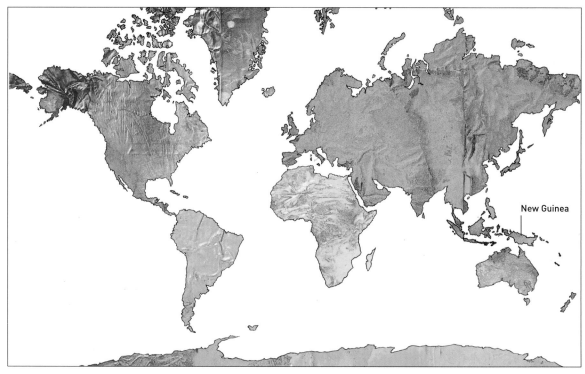

New Guinea

itself – if it could think – would think that no time at all. Written history goes back only about 5,000 years. Go back a million years (that's 200 times as far back as written history stretches) and the continents all have pretty much the same shapes they do today, as far as our eyes would notice. But go back 100 million years and what do we see?

Just look at the map below! The South Atlantic Ocean is a narrow channel by comparison with today, and it looks as though you could almost swim from Africa to South America. Northern Europe is nearly touching Greenland, which is nearly touching Canada. And look where India is: not part of Asia at all, but right down by Madagascar, and tilted on its side. Africa is lurching over the same way, too, compared with the more upright stance we see today.

Come to think of it, did you ever notice, when looking at a modern map, that the eastern side of South America looks suspiciously like the western side of Africa, as though they 'wanted' to fit together, like pieces in a jigsaw puzzle? It turns out that, if we go back a bit further in time (well, about 50 million years further back, but even that is just 'a bit' on the vast, slow geological timescale), we find that they actually did fit together. The map on the right below shows what the southern continents looked like 150 million years ago.

Africa and South America were completely joined up, not just to each other but to Madagascar, India and Antarctica too – and to Australia and New Zealand, round the other side of Antarctica, although you can't see that in the picture. They were all one big land mass called Gondwana (well, it wasn't called Gondwana then – the dinosaurs who lived there didn't call anything anything, but we call it Gondwana today). Gondwana later split up into pieces, creating one daughter continent after another.

It sounds like a pretty tall story, doesn't it? I mean, it sounds pretty ridiculous that anything as massive as a continent could move thousands of miles – but we now know that it happened, and what is more, we understand how.

100 million years ago ▼

150 million years ago ▼

How the Earth moves

We also know that the continents don't only move away from each other. Sometimes they bump into each other, and when that happens huge mountain ranges get pushed up towards the sky. That's how the Himalayas were formed: when India collided with Asia. Actually, it isn't quite true that India collided with Asia. As we shall see soon, what collided with Asia was a much bigger thing, called a 'plate', with India sitting on top of it. All continents sit on these 'plates'. We'll come to them soon, but first let's think a bit more about these 'collisions', and about the continents moving apart.

When you hear a word like 'collided' you might think of a sudden crash, as when a truck collides with a car. That isn't the way it was – and is. The movement of the continents happens agonizingly slowly. Somebody once said it happens about as fast as fingernails grow. If you sit and stare at your fingernails, you don't see them growing. But if you wait a few weeks, you can see that they have grown, and you need to cut them. In the same way, you can't see South America in the act of moving away from Africa. But if you wait 50 million years, you notice that the two continents have moved a long way apart.

'The speed with which fingernails grow' is the average speed at which the continents move. But fingernails grow at a pretty constant speed, whereas the continents move in jerks: there's a jerk, then a pause of a hundred years or so while the pressure to move again builds up, then another jerk, and so on.

Perhaps now you are beginning to guess what earthquakes really are? That's right: an earthquake is what we feel when one of those jerks happens.

I'm telling you this as a known fact, but how do we know it? And when did we first discover it? That's a fascinating story, which I now need to tell.

Various people in the past have noticed the jigsawy kind of fit between South America and Africa, but they didn't know what to make of it.

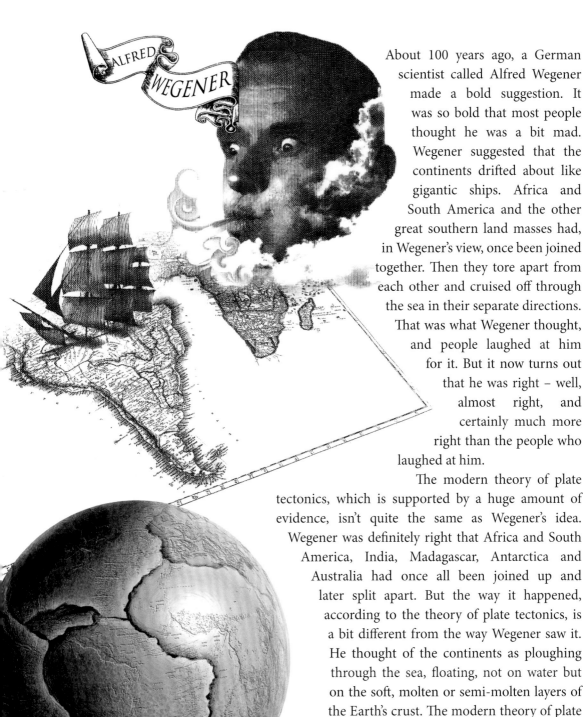

About 100 years ago, a German scientist called Alfred Wegener made a bold suggestion. It was so bold that most people thought he was a bit mad. Wegener suggested that the continents drifted about like gigantic ships. Africa and South America and the other great southern land masses had, in Wegener's view, once been joined together. Then they tore apart from each other and cruised off through the sea in their separate directions. That was what Wegener thought, and people laughed at him for it. But it now turns out that he was right – well, almost right, and certainly much more right than the people who laughed at him.

The modern theory of plate tectonics, which is supported by a huge amount of evidence, isn't quite the same as Wegener's idea. Wegener was definitely right that Africa and South America, India, Madagascar, Antarctica and Australia had once all been joined up and later split apart. But the way it happened, according to the theory of plate tectonics, is a bit different from the way Wegener saw it. He thought of the continents as ploughing through the sea, floating, not on water but on the soft, molten or semi-molten layers of the Earth's crust. The modern theory of plate tectonics sees the whole crust of the Earth, including the bottom of the sea, as a complete set of interlocking plates. (This is 'plates' as in 'armour plates', not the kind of plates you eat off.) So it isn't just the continents that move: it's the plates that they sit on, and there is no bit of the Earth's surface that isn't part of a plate.

Most of the area of most of the plates lies under the sea. The land masses we know as the continents are the high ground of the plates, sticking up above the water. Africa is just the top of the much larger African plate, which stretches halfway across the South Atlantic. South America is the top of the South American plate, which stretches across the other half of the South Atlantic. Other plates are the Indian and Australian plates; the Eurasian plate, which consists of Europe and all of Asia except India; the Arabian plate, which is rather small and slots in between the Eurasian plate and the African plate; and the North American plate, which includes Greenland as well as North America and reaches halfway across the bottom of the North Atlantic ocean. And there are some plates that have hardly any dry land on them at all, for example the vast Pacific plate.

NORTH AMERICAN PLATE

JUAN DE FUCA PLATE

CARIBBEAN PLATE

COCOS PLATE

NAZCA PLATE

SOUTH AMERICAN PLATE

SCOTIA PLAT

PACIFIC PLATE

ANTARCTIC PLATE

EURASIAN PLATE

ARABIAN
PLATE

INDIAN
PLATE

PHILIPPINE
PLATE

AFRICAN PLATE

AUSTRALIAN PLATE

219

You can see from the picture here that the divide between the South American plate and the African plate runs right down the middle of the South Atlantic, miles from either continent. Remember that the plates include the bottom of the sea, and that means hard rock. So how could South America and Africa have nestled together 150 million years ago? Wegener would have had no problem here, because he thought the continents themselves drifted about. But if South America and Africa once snuggled together, how does plate tectonics explain all the undersea hard rock that nowadays separates them? Have the undersea parts of the rocky plates somehow managed to grow?

SOUTH AMERICAN PLATE

AFRICAN PLATE

MID-ATLANTIC RIDGE

SOUTH
AMERICA

SOUTH AMERICAN PLATE
◄MOVING BELT

Sea-floor spreading

Yes. The answer lies in something called 'sea-floor spreading'. You know those moving walkways that you see at large airports to help people with luggage cover the long distances between, say, the entrance to the terminal and the departure lounge? Instead of having to walk all the way, they step on a moving belt and are carried along to some point where they have to start walking again. The moving walkway at an airport is only just wide enough for two people to stand side by side. But now imagine a moving walkway that is thousands of miles wide, stretching most of the way from the Arctic to the Antarctic. And imagine that, instead of moving at walking pace, it moves at the speed with which fingernails grow. Yes, you've guessed it. South America, and the whole South American plate, is being carried away from Africa and the African plate, on something like a moving walkway that lies deep under the sea bed and stretches from the far north to the far south of the Atlantic Ocean, moving very slowly.

What about Africa? Why isn't the African plate moving in the same direction, and why doesn't it keep up with the South American plate?

The answer is that Africa is on a different moving walkway, one that is travelling in the opposite direction. The African moving walkway goes from west to east, while the South American moving walkway goes from east to west. So what is going on in the middle? Next time you are at a big airport, stop just before you step on the moving walkway and watch it. It wells up out of a slit in the floor, and moves away from you. It is a belt, going round and round, travelling forwards above the floor and coming back towards you under the floor. Now imagine another belt, welling out of the same slit but going in exactly the opposite direction. If you put one foot on one belt and the other foot on the other belt you'd be forced to do the splits.

The equivalent of the slit in the floor at the bottom of the Atlantic Ocean runs all along the deep sea floor from the far south to the far north. It is called the mid-Atlantic ridge.

The two 'belts' well up through the mid-Atlantic ridge and head off in opposite directions, one carrying South America steadily westwards, the other carrying Africa away to the east. And, like the belts at the airport, the great belts that move the tectonic plates roll around and come back deep within the Earth.

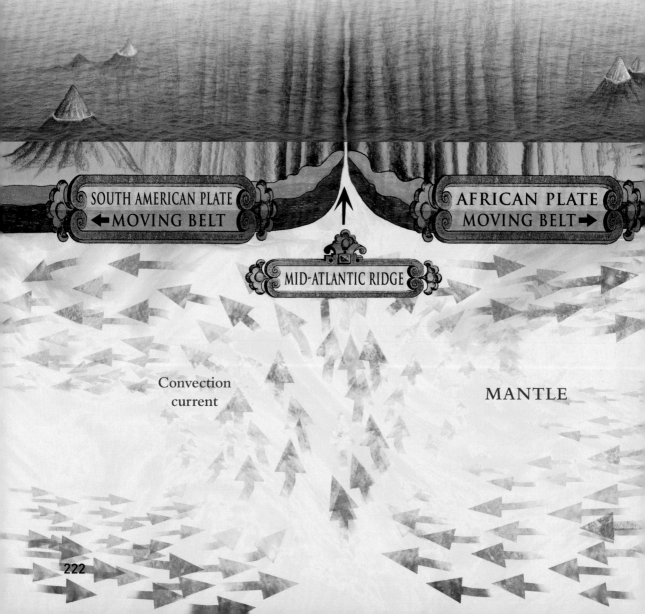

SOUTH AMERICAN PLATE
← MOVING BELT

AFRICAN PLATE
MOVING BELT →

MID-ATLANTIC RIDGE

Convection current

MANTLE

Next time you are at an airport, get on the moving walkway and let it carry you, while you imagine you are Africa (or South America if you prefer). When you get to the other end of the walkway and step off, watch the belt dive underground, ready to make its way back to where you've just come from.

The moving belts at an airport are driven by electric motors. What drives the moving belts that carry the great plates of the Earth with their cargo of continents? Deep beneath the Earth's surface there are what are called convection currents. What's a convection current? Maybe you have an electric convector heater in your house. Here's how it works to heat a room. It heats air. Hot air rises because it is less dense than cold air (that's how hot-air balloons work). The hot air rises until it hits the ceiling, where it can't rise any more and is forced sideways by the fresh hot air pushing up from

AFRICA

Convection
current

beneath. As it travels sideways, the air cools down, whereupon it sinks. When it hits the floor, it again moves sideways, creeping along the floor until it gets caught up in the heater and rises again. That explanation is a bit too simple, but the basic idea is all that matters here: under ideal conditions a convector heater can get the air moving round and round – circulating. This kind of circulation is called a 'convection current'.

The same thing happens in water. In fact, it can happen in any liquid or any gas. But

how can there be convection currents under the Earth's surface? It isn't liquid down there, is it? Well, yes, it is – sort of. Not liquid like water, but sort of half liquid like thick honey or treacle. That's because it is so hot that everything is melting. The heat comes from deep down. The centre of the Earth is very hot indeed, and it goes on being hot until much closer to the surface. Occasionally the heat bursts out through the surface at a place we call a volcano.

The plates are made of hard rock, and, as we've seen, most of them is under the sea. Each plate is several miles thick. This thick layer of armour plating is called the lithosphere, which literally means 'sphere of rock'. Under the sphere of rock is an even thicker layer, if you can believe it, which isn't actually called the sphere of treacle but probably should be (it's actually the upper mantle). The hard rocky plates of the sphere of rock could be said to 'float' on the sphere of treacle. Deep heat beneath and within the sphere of treacle causes agonizingly slow, grinding convection currents in the treacle, and it is those convection currents that carry the great rocky plates floating above.

Convection currents follow pretty complicated paths. Just think about all the different ocean currents, and even the winds, which are sort of high-speed convection currents. So it's no wonder that the various plates on the Earth's surface are carried in all sorts of directions, rather than round and round as if they were all on a simple merry-go-round. No wonder the plates

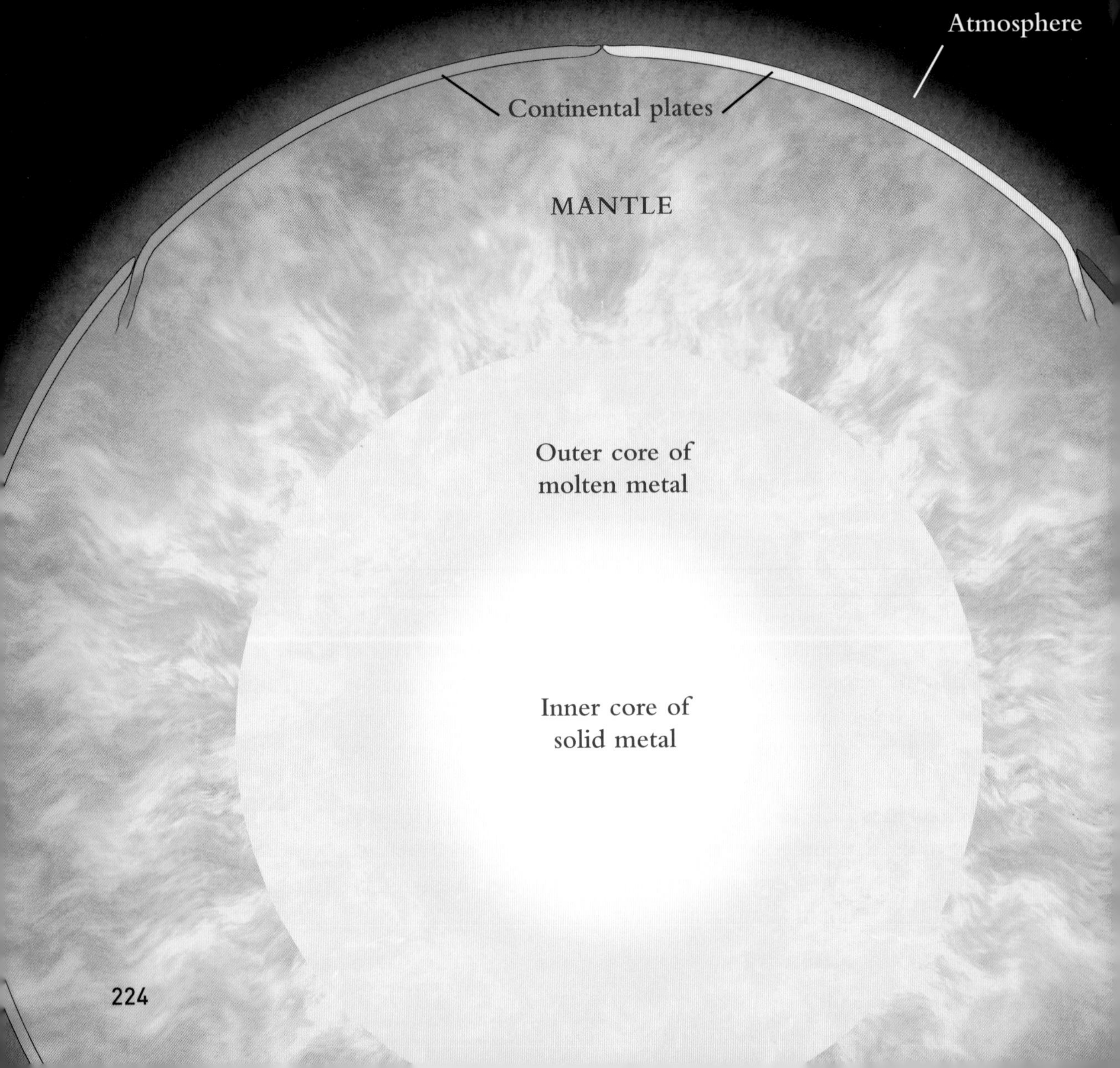

Atmosphere

Continental plates

MANTLE

Outer core of
molten metal

Inner core of
solid metal

bump into each other or tear rendingly away from each other, dive one under the other or grate sideways against each other. And no wonder we feel these titanic forces – grinding, wrenching, roaring, scraping forces – as earthquakes. Terrible as earthquakes can be, the wonder is that they aren't even more terrible.

Sometimes a moving plate slides underneath a neighbouring plate. This is called 'subduction'. Part of the African plate, for example, is being subducted under the Eurasian plate. This is one reason why there are earthquakes in Italy, and it is one reason why Mount Vesuvius erupted in ancient Roman times and destroyed the towns of Pompeii and Herculaneum (because volcanoes tend to sprout along the edges of the plates). The Himalayan mountains, including Mount Everest, were forced up to their great height as the Indian plate was steadily subducted under the Eurasian plate.

We began with the San Andreas Fault, so let's end there. The San Andreas Fault is a long, rather straight 'slippage' line between the Pacific plate and the North American plate. Both plates are moving north-west, but the Pacific plate is moving faster. The city of Los Angeles lies on the Pacific plate, not the North American plate, and is steadily creeping up on San Francisco, most of which is on the North American plate. Earthquakes are constantly to be expected in this whole region, and experts are predicting that there will be a big one within the next ten years or so. Fortunately, California, unlike Haiti, is well equipped to deal with the terrible plight of earthquake victims.

One day, parts of Los Angeles might end up in San Francisco. But that is a long way off, and none of us will be around to see it.

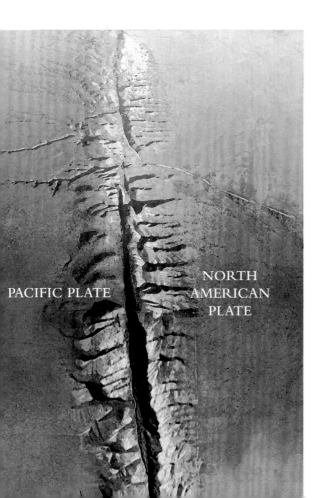

PACIFIC PLATE

NORTH AMERICAN PLATE

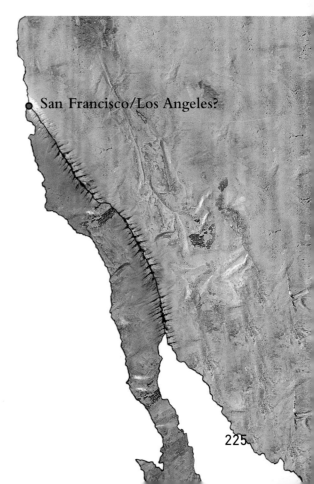

San Francisco/Los Angeles?

225

11

WHY DO BAD THINGS HAPPEN?

WHY DO bad things happen? After a dreadful disaster such as an earthquake or a hurricane you'll hear people saying things like this:

'It's so unfair. What did those poor people ever do to deserve such a fate?'

If a really good person gets a painful disease and dies, while a really bad person remains in the best of health, once again we cry,

'Unfair!' Or we say,

'Where's the justice in that?'

It is hard to resist this feeling that, somehow, there ought to be a kind of natural justice. Good things should happen to good people. Bad things, if they must happen at all, should only happen to bad people. In Oscar Wilde's delightful play *The Importance of Being Earnest,* an elderly governess called Miss Prism explains how, long ago, she wrote a novel. When she is asked whether it ended happily, she replies: 'The good ended happily, and the bad unhappily. That is what fiction means.' Real life is different. Bad things

226

do happen, and they happen to good people as well as bad. Why? Why is real life not like Miss Prism's fiction? Why do bad things happen?

Lots of peoples believe that their gods intended to create a perfect world but unfortunately something went wrong – and there are almost as many ideas about what that something was. The Dogon tribe of West Africa believe that at the beginning of the world there was a cosmic egg from which two twins hatched. All would have been well if the twins had hatched at the same time. Unfortunately, one of them hatched too soon, and spoiled the gods' plan of perfection. That, according to the Dogon, is why bad things happen.

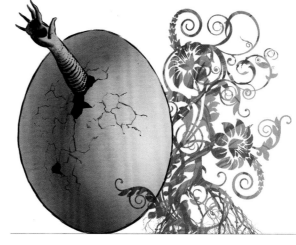

There are lots of legends about how death came into the world. All over Africa, different tribes believe that the chameleon was given the news of everlasting life and told to carry it to humans. Unfortunately the chameleon walked so slowly (they do, I know: as a child in Africa I had a pet chameleon called Hookariah) that the news of death, carried by a nippier lizard (or other faster animal in other versions of the legend), arrived first. In one West African legend, the news of life was brought by a slow toad, unfortunately overtaken by a fast dog bringing the news of death. I must say I'm a bit puzzled why *the order in which news arrives* should matter so much. Bad news is still bad, whenever it arrives.

227

Disease is a special kind of bad thing, and it has spawned plenty of myths of its own. One reason is that for a long time diseases were rather mysterious. Our ancestors faced other dangers – from lions and sabretooths, from enemy tribes, from the threat of starvation – but you could see them coming, and understand them. Smallpox, on the other hand, or the Black Death, or malaria, must have seemed to pounce from nowhere, without warning, and it wasn't obvious how to guard against these assaults. It was a terrifying mystery. Where did diseases come from? What did we do to deserve this painful death, this agonizing toothache or these hideous spots? No wonder people resorted to superstition when desperately trying to understand disease, and even more desperately trying to protect themselves from it. In many African tribes, until quite recently, anybody who got ill, or had a sick child, would automatically look around for an evil magician or witch to blame.

If my child has a high fever, it must be because an enemy paid a witch doctor to cast a spell on her. Or maybe it is because I couldn't afford to sacrifice a goat when she was born. Or perhaps it is because a green caterpillar walked across the path in front of me and I forgot to spit out the evil spirit.

In ancient Greece, sick pilgrims would spend the night in a temple dedicated to Asclepius, the god of healing and medicine. They believed the god would either heal them himself or reveal the cure in a dream. Even today, a surprisingly large number of sick people travel to places like Lourdes, where they plunge into a sacred pool in the hope that the holy water will heal them (actually, I suspect that they are more likely to catch something from all the other people who have bathed in the same water). About 200 million people have made the pilgrimage to Lourdes during the past 140 years, hoping for a cure. In many cases there is not much wrong with them, and thankfully they

mostly get better – as they would have anyway, with or without the pilgrimage.

Hippocrates, the ancient Greek 'father of medicine' who gives his name to the oath of good conduct that all doctors are supposed to observe, thought that earthquakes were important causes of disease. In the middle ages, many people believed that diseases were caused by the movements of the planets against the backdrop of stars. That's part of a system of beliefs called astrology, which, ridiculous as it may seem, still has quite a few followers to this day.

The most persistent myth about health and disease, lasting from the fifth century BC right up to the eighteenth century AD, was the myth of the four 'humours'. When we say, 'He's in a good humour today,' that's where the word comes from, although people don't believe in the idea behind it any more. The four humours were black bile, yellow bile, blood and phlegm. Good health was thought to depend on a good 'balance' between them, and you can still hear something a bit similar from quack 'healers' today who will wave their hands over you in order to 'balance' your 'energies' or your 'chakras'.

The theory of the four humours certainly couldn't help doctors to cure illnesses, but it might have done no great harm except that it led to the practice of 'bleeding' patients. This involved opening a vein with a sharp instrument called a lancet, and drawing off quantities of blood into a special basin. This, of course, made the poor patient even sicker (it contributed to George Washington's death) – but the doctors believed so strongly in the ancient myth of the humours that they did it again and again. What's more, people didn't only get bled when they were ill. Sometimes they asked the doctor to do it in advance of getting ill, in the hope that it would ward off sickness.

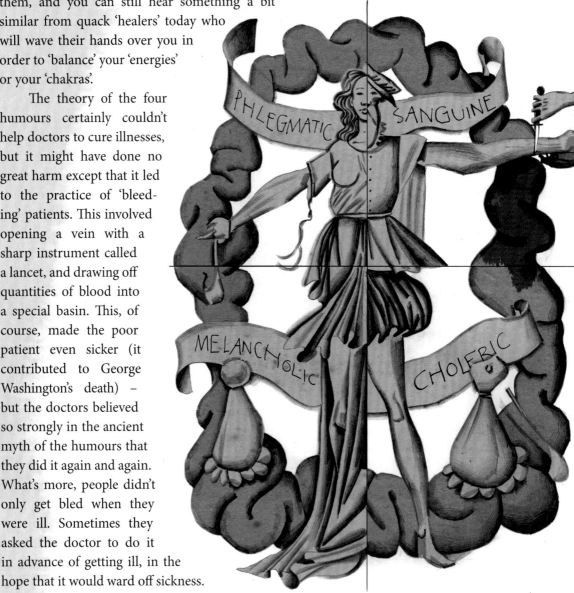

Once, when I was at school, our teacher asked us to think about why diseases happen. One boy put his hand up and suggested that it was because of 'sin'! There are many people, even today, who think something like that is the cause of bad things generally. Some myths suggest that bad things happen in the world because our ancestors did something wicked long ago. I've already mentioned the

Jewish myth of the founding ancestors Adam and Eve. You'll remember that Adam and Eve did a simply terrible thing: they allowed themselves to be persuaded by the snake to eat the fruit of a forbidden tree. This mythical crime has reverberated down the ages and is still regarded by some people as responsible for all the bad things that happen in the world to this day.

Lots of myths talk about a conflict between good gods and bad gods (or devils). The bad gods are responsible for the bad things that happen in the world. Or there may be a single spirit of evil, called the Devil or something similar, who fights with the good god or gods. If only there wasn't this tussle between devils and gods, or good gods and bad gods, bad things wouldn't happen.

Why do **bad things** happen *really*?

WHY DOES *anything* happen? That's a complicated question to answer, but it is a more sensible question than 'Why do *bad* things happen?' This is because there is no reason to single out bad things for special attention unless bad things happen more often than we would expect them to, by chance; or unless we think there should be a kind of natural justice, which would mean that bad things should only happen to bad people.

Do bad things happen more often than we ought to expect by chance alone? If so, then we really do have something to explain. You may have heard people refer jokingly to 'Murphy's Law', sometimes called 'Sod's Law'. This states: 'If you drop a piece of toast and marmalade on the floor, it always lands marmalade side down.' Or, more generally: 'If a thing can go wrong, it will.' People often joke about this, but at times you get the feeling they think it is more than a joke. They really do seem to believe the world is out to get them.

I do a certain amount of filming for television documentaries, and one of the things that can go wrong in filming 'on location' is unwanted noise. When an aircraft drones in the distance, you have to stop filming and wait for it to go, and this can be extremely irritating. Costume dramas of life in earlier centuries are ruined by even a trace of aircraft noise. Film people have a superstition that aircraft deliberately choose moments when silence is most important to fly overhead, and they invoke Sod's Law.

Recently, a film crew I was working with chose a location where we felt sure there should be a minimum of noise, a huge empty meadow near Oxford. We arrived early in the morning to make doubly sure of peace and quiet – only to discover, when we arrived, a lone Scotsman practising the bagpipes (perhaps banished from the house by his wife). 'Sod's Law!' we all proclaimed. The truth, of course, is that there is noise going on most of the time, but we only *notice* it when it is an irritation, for example when it interferes with filming. There is a bias in our likelihood of noticing annoyance, and this makes us think the world is out to annoy us deliberately.

In the case of the toast, it wouldn't be surprising to find that it really does fall marmalade side down more often than not, because tables are not very high, the toast starts marmalade side up and there is usually time for one half-rotation before it hits the ground. But the toast example is just a colourful way to express the gloomy idea that

'if a thing can go wrong it will.'

Perhaps this would be a better example of Sod's Law: 'When you toss a coin, the more strongly you want heads, the more likely it is to come up tails.'

That, at least, is the pessimistic view. There are optimists who think that the more you want heads, the more likely the coin is to come up heads. Perhaps we could call that 'Pollyanna's Law' – the optimistic belief that things usually turn out for the good. Or it could be called 'Pangloss's Law', after a character invented by the great French writer Voltaire. His 'Dr Pangloss' thought that 'All is for the best in this best of all possible worlds.'

When you put it like that, you can quickly see that Sod's Law and Pollyanna's Law are both nonsense. Coins, and slices of toast, have no way of knowing the strength of your desires, and no desire of their own to thwart them – or fulfil them. Also, what is a bad thing for one person may be a good thing for another. Rival tennis players may both pray fervently for victory, but one has to lose! There is no special reason to ask, 'Why do bad things happen?' Or, for that matter, 'Why do good things happen?' The real question underlying both is the more general question: 'Why does *anything* happen?'

233

Luck, chance and cause

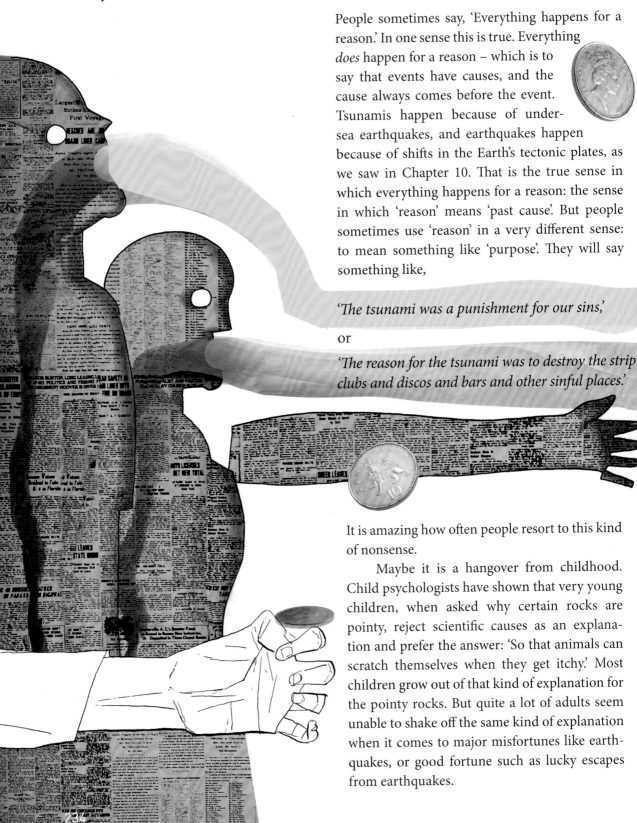

People sometimes say, 'Everything happens for a reason.' In one sense this is true. Everything *does* happen for a reason – which is to say that events have causes, and the cause always comes before the event. Tsunamis happen because of under-sea earthquakes, and earthquakes happen because of shifts in the Earth's tectonic plates, as we saw in Chapter 10. That is the true sense in which everything happens for a reason: the sense in which 'reason' means 'past cause'. But people sometimes use 'reason' in a very different sense: to mean something like 'purpose'. They will say something like,

'The tsunami was a punishment for our sins,'

or

'The reason for the tsunami was to destroy the strip clubs and discos and bars and other sinful places.'

It is amazing how often people resort to this kind of nonsense.

Maybe it is a hangover from childhood. Child psychologists have shown that very young children, when asked why certain rocks are pointy, reject scientific causes as an explanation and prefer the answer: 'So that animals can scratch themselves when they get itchy.' Most children grow out of that kind of explanation for the pointy rocks. But quite a lot of adults seem unable to shake off the same kind of explanation when it comes to major misfortunes like earthquakes, or good fortune such as lucky escapes from earthquakes.

234

What about 'bad luck'? Is there such a thing as bad luck, or indeed good luck? Are some people luckier than others? People sometimes talk of a 'run' of bad luck. Or they will say, 'So many bad things have happened to me lately, I'm due for a piece of really good luck.' Or they may say, 'So-and-so is such an unlucky person, things always seem to turn out badly for her.'

'I'm due for a piece of good luck' is an example of a widespread misunderstanding of the 'Law of Averages'. In the game of cricket, it often makes a big difference which team bats first. The two captains toss a coin to decide who gets the advantage, and each team's supporters very much hope their captain will win the toss. Before a recent match between India and Sri Lanka, a Yahoo web page posed the question:

> *'Will Dhoni [the Indian captain] be* lucky *once again with the toss?'*

Of the answers they received, the following was chosen (for some reason that I don't understand) as 'Best Answer':

> *'I firmly believe in the law of averages, so my bet is on Sangakkara [the Sri Lanka captain] being* lucky *and winning the much hyped toss.'*

Can you see what rubbish this is? In a series of previous matches, Dhoni had won the toss every time. Coins are supposed to be unbiased. So the misunderstood 'Law of Averages' ought to see to it that Dhoni, having been lucky so far, should now lose the toss, to *redress the balance*. Another way to put this would be to say that it was now Sangakkara's *turn* to win the toss. Or that it would be *unfair* if Dhoni won the toss yet again. But the reality is that, no matter how many times Dhoni has won the toss before, the chances that he will win it again this time are *always* 50:50. 'Turns' and 'fairness' simply don't come into it. *We* may care about fairness and unfairness, but coins don't give a toss! Nor does the universe at large.

It is true that if you toss a penny 1,000 times, you'd expect approximately 500 heads and 500 tails. But suppose you've tossed the penny 999 times and it's so far come up heads every time. What would you bet for the last toss? According to the widespread misunderstanding of the 'Law of Averages', you should bet on tails, because it is tails' *turn*, and it would be so *unfair* if it came up heads yet again. But I would place my bet on heads, and so would you if you were wise. A sequence of 999 heads in a row suggests that someone has tinkered with the penny, or with the method of tossing it. The misunderstood 'Law of Averages' has been the ruin of many gamblers.

Admittedly, with hindsight you can say, 'Sangakkara was very unlucky to lose the toss, because it meant that India batted on a perfect pitch and that helped them to amass a huge score.' There is nothing wrong with that. All you are saying is that this time around winning the toss really made a difference, so whoever won the toss on this particular occasion was very lucky to have done so. What you should *not* say is that because Dhoni has won the toss on many occasions before, it is Sangakkara's turn this time! Nor should you ever say something like this: 'Dhoni happens to be a good cricketer, but the real reason we should make him captain is that he is very lucky at winning the toss.' Luck with coin tosses is not something that individual people possess. You can say of a cricketer that he is a good batsman or a bad bowler. You can*not* say that he is good at winning tosses, or bad at winning tosses!

For just the same reason, it is complete nonsense to think you can improve your luck by wearing a lucky charm around your neck. Or by crossing your fingers behind your back. These things have no way of influencing what happens to you unless it is by some effect on how you feel: giving you added confidence that calms your nerves before a tennis serve, for example. But that is nothing to do with luck; that is psychology.

True, some people are described as 'accident prone'. This is fine, if it only means something like 'clumsy', or especially likely to fall over or otherwise suffer misfortune.

If you want a really funny example of 'accident prone', see the hilarious film *The Pink Panther*, starring Peter Sellers as Inspector Jacques Clouseau. Inspector Clouseau continually has embarrassing and amusing accidents, but that is because he is a habitual bungler, not because he has constant bad 'luck', which is how some people use the phrase.

(By the way, do try to see the original *Pink Panther* film itself, not the later run of inferior films with similar titles like *Son of Pink Panther*, *The Pink Panther's Revenge* and so on, which it spawned.)

Pollyanna and paranoia

So, we have seen that bad things, like good things, don't happen any more often than they ought to by chance. The universe has no mind, no feelings and no personality, so it doesn't do things in order to either hurt or please you. Bad things happen because *things* happen. Whether they are bad or good from our point of view doesn't influence how likely it is that they will happen. Some people find it hard to accept this. They'd prefer to think that sinners get their come-uppance, that virtue is rewarded. Unfortunately the universe doesn't care what people prefer.

But now, having said all that, I pause for thought. Funnily enough, I have to admit that something a bit like Sod's Law is true. Although it is definitely not true that the weather, or an earthquake, is out to get you (for they don't care about you, one way or the other) things are a bit different when we turn to the living world. If you are a rabbit, a fox is out to get you. If you are a minnow, a pike is out to get you. I don't mean

the fox or pike thinks about it, although it may. I'd be equally happy to say that a virus is out to get you, and nobody believes viruses think about anything. But evolution by natural selection has seen to it that viruses, and foxes, and pikes, behave in ways that are actively bad for their victims – behave as though they are deliberately out to get them – in ways that you couldn't say of earthquakes or hurricanes or avalanches. Earthquakes and hurricanes are bad for their victims, but they don't take active steps to do bad things: they don't take active steps to do anything, they just happen.

Natural selection, the struggle for existence as Darwin called it, means that every living creature has enemies that are working hard for its downfall. And sometimes the tricks that natural enemies use give the appearance of being cleverly planned. Spider webs, for example, are ingenious traps laid for unsuspecting insects. A little insect called an ant lion digs booby traps for its prey to fall into.

239

The ant lion itself sits under the sand at the bottom of the conical pit that it digs, and seizes any ant that falls into the pit. Nobody is suggesting that the spider or the ant lion is ingenious – that it has *thought up* its cunning trap. But natural selection makes them evolve brains that behave in ways that *look* ingenious to our eyes. In the same way, a lion's body looks ingeniously designed to bring about the doom of antelopes and zebras. And we can imagine that, if you were an antelope, a stalking, chasing, pouncing lion might seem out to get you.

It's easy to see that predators (animals that kill and then eat other animals) are working for the downfall of their prey. But it's also true that prey are working for the downfall of their predators. They work hard to escape being eaten, and if they all succeeded the predators would starve to death. The same thing holds between parasites and their hosts. It also holds between members of the same species, all of whom are actually or potentially competing with one another. If the living is easy, natural selection will favour the evolution of improvements in enemies, whether predators, prey, parasites, hosts

or competitors: improvements that will make life hard again. Earthquakes and tornadoes are unpleasant and might even be called enemies, but they are not 'out to get you' in the same 'Sod's Law' kind of way that predators and parasites are.

This has consequences for the sort of mental attitude that any wild animal, such as an antelope, might be expected to have. If you are an antelope and you see the long grass rustling, it could be just the wind. That's nothing to worry about, for the wind is not out to get you: it is completely indifferent to antelopes and their well-being. But that rustle in the long grass could be a stalking leopard, and a leopard is most definitely out to get you: you taste good to a leopard and natural selection favoured ancestral leopards that were good at catching antelopes. So antelopes and rabbits and minnows, and most other animals, have to be constantly on the alert. The world is full of dangerous predators and it is safest to assume that something a bit like Sod's Law is true. Let's put that in the language of Charles Darwin, the language of natural selection: those individual animals that

act as though Sod's Law were true are more likely to survive and reproduce than those individual animals that follow Pollyanna's Law.

Our ancestors spent much of their time in mortal danger from lions and crocodiles, pythons and sabretooths. So it probably made sense for each person to take a suspicious – some might even say paranoid – view of the world, to see a likely threat in every rustle of the grass, every snap of a twig, and to assume that something was out to get him, a deliberate agent scheming to kill him. 'Scheming' is the wrong way to look at it if you think of it as deliberate plotting, but it is easy to put the idea into the language of natural selection: 'There are enemies out there, shaped by natural selection to behave as though they were scheming to kill me. The world is not neutral and indifferent to my welfare. The world is out to get me. Sod's Law may or may not be true, but behaving as if it is true is safer than behaving *as if* Pollyanna's Law is true.'

Maybe this is one reason why, to this day, many people have superstitious beliefs that the world is out to get them. When this goes too far, we say they are 'paranoid'.

Illness and evolution – work in progress?

As I said, predators aren't the only things that are out to get us. Parasites are a more sneaky threat, but they are just as dangerous. Parasites include tapeworms and flukes, bacteria and viruses, which make a living by feeding off our bodies. Predators such as lions also feed off bodies, but the distinction between a predator and a parasite is usually clear. Parasites feed off still-living victims (though they may eventually kill them) and they are usually smaller than their victims. Predators are either larger than their victims (as a cat is larger than a mouse) or, if smaller (as a lion is smaller than a zebra), not very much smaller. Predators kill their prey outright and then eat them. Parasites eat their victims more slowly, and the victim may stay alive a long time with the parasite gnawing away inside.

Parasites often attack in large numbers, as when our body suffers a massive infection with a flu or cold virus. Parasites that are too small to see with the naked eye are often called 'germs', but that's rather an imprecise word. They include viruses, which are very very small indeed; bacteria, which are larger than viruses but still very small (there are viruses that act as parasites on bacteria); and other single-celled organisms like the malarial parasite, which are much larger than bacteria but still too small to be seen without a microscope. Ordinary language has no general name for these larger singled-celled parasites. Some of them can be called 'protozoa', but that's now rather an outdated term. Other important parasites include fungi, for example ringworm and athlete's foot (big things like mushrooms and toadstools give a false impression of what most fungi are like).

Examples of bacterial diseases are tuberculosis, some kinds of pneumonia, whooping cough, cholera, diphtheria, leprosy, scarlet fever, boils and typhus. Viral diseases include measles, chickenpox, mumps, smallpox, herpes, rabies, polio, rubella, various varieties of influenza and the cluster of diseases that we call the 'common cold'. Malaria, amoebic dysentery and sleeping sickness are among those diseases caused by 'protozoa'. Other important parasites, larger still – large enough to be seen with the naked eye – are the various kinds of worms, including flatworms, roundworms and flukes. When I was a boy living on a farm, I would quite often find a dead

animal like a weasel or a mole. I was learning biology at school, and I was interested enough to dissect these little corpses when I found them. The main thing that impressed me was how full of wriggling, live worms they were (roundworms, technically called nematodes). The same was never true of the domesticated rats and rabbits we were given to dissect at school.

The body has a very ingenious and usually effective system of natural defence against parasites, called the immune system. The immune system is so complicated that it would take a whole book to explain it. Briefly, when it senses a dangerous parasite the body is mobilized to produce special cells, which are carried by the blood into battle like a kind of army, tailor-made to attack the particular parasites concerned. Usually the immune system wins, and the person recovers. After that, the immune system 'remembers' the molecular equipment that it developed for that particular battle, and any subsequent infection by the same kind of parasite is beaten off so quickly that we don't notice it. That is why, once you have had a disease like measles or mumps or chickenpox, you're unlikely to get it again. People used to think it was a good idea if children caught mumps, say, because the immune system's

'memory' would protect them against getting it as an adult – and mumps is even more unpleasant for adults (especially men, because it attacks the testicles) than it is for children. Vaccination is the ingenious technique of doing something similar on purpose. Instead of giving you the disease itself, the doctor gives you a weaker version of it, or possibly an injection of dead germs, to stimulate the immune system without actually giving you the disease. The weaker version is much less nasty than the real thing: indeed, you often don't notice any effect at all. But the immune system 'remembers' the dead germs, or the infection with the mild version of the disease, and so is forearmed to fight the real thing if it should ever come along.

The immune system has a difficult task 'deciding' what is 'foreign' and therefore to be fought (a 'suspected' parasite), and what it should accept as part of the body itself. This can be particularly tricky, for example, when a woman is pregnant. The baby inside her is 'foreign' (babies are not genetically identical to their mothers because half their genes come from the father). But it is important for the immune system not to fight against the baby. This was one of the difficult problems that had to be solved when pregnancy evolved in the ancestors of mammals. It was solved –

after all, plenty of babies do manage to survive in the womb long enough to be born. But there are also plenty of miscarriages, which perhaps suggests that evolution had a hard time solving it and that the solution isn't quite complete. Even today, many babies survive only because doctors are on hand – for example, to change their blood completely as soon as they are born, in some extreme cases of immune-system overreaction.

Another way in which the immune system can get it wrong is to fight too hard against a supposed 'attacker'. That is what allergies are: the immune system needlessly, wastefully and even damagingly fighting harmless things. For example, pollen in the air is normally harmless, but the immune system of some people overreacts to it – and that's when you get the allergic reaction called 'hay fever': you sneeze and your eyes water, and it is very unpleasant. Some people are allergic to cats, or to dogs: their immune systems are overreacting to harmless molecules in or on the hair of these animals. Allergies can sometimes be very dangerous. A few people are so allergic to peanuts that eating a single one can kill them.

Sometimes an overreacting immune system goes so far that a person is allergic to himself! This causes so-called auto-immune diseases (*autos* is Greek for 'self'). Examples of auto-immune diseases are alopecia (your hair falls out in patches because the body attacks its own hair follicles) and psoriasis (an overactive immune system causes pink scaly patches on the skin).

It is not surprising that the immune system sometimes overreacts, because there's a fine line to be trodden between failing to attack when you should and attacking when you shouldn't. It's the same problem we met over the antelope trying to decide whether to run away from the rustle in the long grass. Is it a leopard? Or is it a harmless puff of wind stirring the grass? Is this a dangerous bacterium, or is it a harmless pollen grain? I can't help wondering whether people with a hyperactive immune system, who pay the penalty of allergies or even auto-immune diseases, might be less likely to suffer from certain kinds of viruses and other parasites.

Such 'balance' problems are all too common. It is possible to be too 'risk averse' – too jumpy, treating every rustle in the grass as danger, or unleashing a massive immune response to a harmless peanut or to the body's own tissues. And it is possible to be too gung-ho, failing to respond to

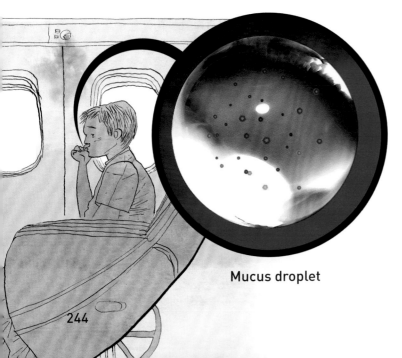

Mucus droplet

How the immune system deals with an attempted attack by a flu virus (right)

The top sequence shows a successful attack. A flu virus approaches a cell (**1**). The virus key matches the cell lock (cell surface receptor) (**2**), so that the virus is allowed into the cell (**3**), where it replicates. Finally (**4**), hundreds of replicated viruses burst out of the infected cell.

The bottom sequence shows the immune system fighting off the attack. Antibodies approach the virus (**1**) and attach themselves to it (**2**). Now the virus key no longer fits the cell lock (**3**), so the virus cannot enter the cell.

danger when it is very real, or failing to mount an immune response when there really is a dangerous parasite. Treading the line is difficult, and there are penalties for straying off it in either direction.

Cancers are a special case of a bad thing that happens: a strange one, but a very important one. A cancer is a group of our own cells that have broken away from doing what they are supposed to do in the body and have become parasitic. Cancer cells are usually grouped together in a 'tumour', which grows out of control, feeding on some part of the body. The worst cancers then spread to other parts of the body (that's called metastasis) and eventually often kill it. Tumours that do this are called malignant.

The reason cancers are so dangerous is that their cells are directly derived from the body's own cells. They are our own cells, slightly modified. This means the immune system has a hard time recognizing them as foreign. It also means it is very difficult to find a treatment that kills the cancer, because any treatment you can think of – like a poison, say – is likely to kill our own healthy cells as well. It is much easier to kill bacteria, because bacterial cells are different from ours. Poisons that kill bacterial cells but not our own cells are called antibiotics. Chemotherapy poisons cancer cells, but it also poisons the rest of us because our cells are so similar. If you overdo the dose of the poison, you may kill the cancer, but not before killing the poor patient.

We're back to the same problem of striking a balance between attacking genuine enemies (cancer cells) and not attacking friends (our own normal cells): back to the problem of the leopard in the long grass again.

Let me end this chapter with a speculation. Is it possible that auto-immune diseases are a kind of byproduct of an evolutionary war, over many ancestral generations, against cancer? The immune system wins many battles against pre-cancerous cells, suppressing them before they have a chance to become fully malignant. My suggestion is that, in its constant vigilance against pre-cancerous cells, the immune system sometimes goes too far and attacks harmless tissues, attacks the body's own cells – and we call this an auto-immune disease. Could it be that the explanation of auto-immune diseases is that they are evidence of evolution's work-in-progress on an effective weapon against cancer?

What do you think?

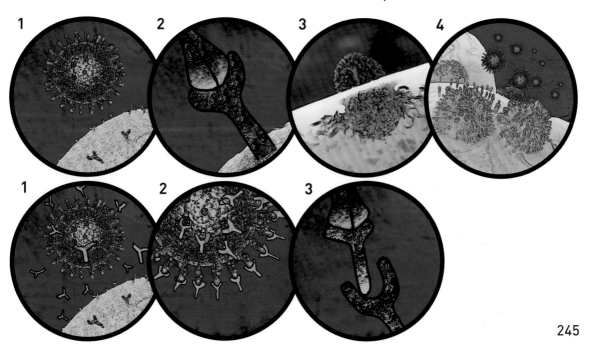

245

MIRACLE?

IN THE FIRST chapter of this book I talked about magic, and separated supernatural magic (casting a spell to turn a frog into a prince, or rubbing a lamp to conjure up a genie) from conjuring tricks (illusions, such as silk handkerchiefs turning into rabbits, or women being sawn in half). Nobody nowadays believes in fairytale magic. Every-body knows that pumpkins turn into coaches only in *Cinderella*. And we all know that rabbits come out of apparently empty hats only by trickery. But there are some supernatural tales that are still taken seriously, and the 'events' they recount are often called miracles. This chapter is about miracles – stories of supernatural happenings that many people believe, as opposed to fairytale spells, which nobody believes, and conjuring tricks, which look like magic but we know are faked.

Some of these tales are ghost stories, spooky urban legends or stories of uncanny coincidence – stories like, 'I dreamed about a celebrity whom I hadn't thought about for years, and the very next morning I heard that he'd died in the night.' Many more come from the hundreds of religions around the world, and these in particular are often called miracles. To take just one example, there is a legend that, about 2,000 years ago, a wandering Jewish preacher called Jesus was at a wedding where they ran out of wine. So he called for some water and used miraculous powers to turn it into wine – very good wine, as the story goes on to tell us. People who would laugh at the idea that a pumpkin could turn into a coach, and who know perfectly well that silk handkerchiefs don't really turn into rabbits, are quite happy to believe that a prophet turned water into wine or, as devotees of another religion would have it, flew to heaven on a winged horse.

Rumour, coincidence and snow-balling stories

Usually when we hear a miracle story it's not from an eye witness, but from somebody who heard about it from somebody else, who heard it from somebody else, who heard it from somebody else's wife's friend's cousin . . . and any story, passed on by enough people, gets garbled. The original source of the story is often itself a rumour that began so long ago and has become so distorted in the retelling that it is almost impossible to guess what actual event – if any – started it off.

After the death of almost any famous person, hero or villain, stories that somebody has seen them alive start rushing around the globe. This was true of Elvis Presley, of Marilyn Monroe, even of Adolf Hitler. It's hard to know why people enjoy passing on such rumours when they hear them, but the fact is that they do, and that is a big part of the reason why rumours spread.

Here's a recent example of how such a rumour gets started. Soon after the popular singer Michael Jackson died in 2009, an American television crew was given a guided tour of his famous mansion called Neverland.

In one scene of the resulting film, people thought they saw his ghost at the end of a long corridor. I've looked at a recording, and it is very unconvincing; however, it was enough to start wild rumours flying around. Michael Jackson's ghost is at large! Copycat sightings soon emerged. For example, on the opposite page is a photograph that a man took of the polished surface of his car. To you and me, especially when we compare the 'face' with the other clouds on either side, what we are looking at is obviously the reflection of a cloud. But to the overheated imagination of the devoted fan it could only be the ghost of Michael Jackson, and the picture on YouTube has received more than 15 million hits!

Actually, there's something interesting going on here, which is worth mentioning. Humans are social animals, so the human brain is pre-programmed to see the faces of other humans, even where there aren't any. This is why people so often imagine they see faces in the random patterns made by clouds, or on slices of toast, or in damp patches on walls.

Spine-tingling ghost stories are fun to tell, especially if they are really scary, and even more so if you claim that they are true. When I was eight, my family lived briefly in a house called Cuckoos, about 400 years old, with wonky black Tudor beams. Not surprisingly, the house had a legend about a long-dead priest hidden in a secret passage. There was a story that you could hear his footsteps on the stairs, but with the twist that you could hear one step too many – spookily explained by the fact that the staircase was said to have had an extra step in the sixteenth century! I remember the pleasure I took in passing the story on to my schoolfriends. It never occurred to me to ask how good the evidence was. It was enough that the house was old, and my friends were impressed.

People get a thrill from passing on ghost stories. The same applies to miracle stories. If a rumour of a miracle gets written down in a book, the rumour becomes hard to challenge, especially if the book is ancient. If a rumour is

old enough, it starts to be called a 'tradition' instead, and then people believe it all the more. This is rather odd, because you might think they would realize that older rumours have had more time to get distorted than younger rumours that are close in time to the alleged events themselves. Elvis Presley and Michael Jackson lived too recently for traditions to have grown up, so not many people believe stories like 'Elvis seen on Mars'. But maybe in 2,000 years' time . . .?

What about those strange stories people tell of having a dream about somebody they haven't seen or thought of for years, then waking up to find a letter from that person waiting on the doormat? Or waking up to hear or read that the person died in the night? You may have had such an experience yourself. How do we explain coincidences like that?

Well, the most likely explanation is that they really are just that: coincidences, and nothing more. The key point is that we only bother to tell stories when strange coincidences happen – not when they don't. Nobody ever says, 'Last night

I dreamed about that uncle I haven't thought of for years, and then I woke up and found that *he hadn't died in the night!*'

The more spooky the coincidence, the more likely the news of it will spread. Sometimes it strikes a person as so remarkable that he fires off a letter to a newspaper. Perhaps he dreams, for the first time ever, of a once famous but long forgotten actress from the distant past, then wakes to discover that she died in the night. A 'farewell visit' in a dream – how spooky! But just think for a moment what has actually happened. For a coincidence to be reported in a newspaper, it only has to be experienced by one person among the millions of readers who might write to the paper. If we just take Britain alone, about 2,000 people die every day, and there must be a hundred million dreams every night. When you think of it like that, we'd positively expect that from time to time somebody will wake up and discover that the person they had been dreaming of had died in the night. They are the only ones who would send their stories to the papers.

Another thing that happens is that stories grow in the telling and re-telling. People enjoy a good story so much that they embellish it to make it a bit better than it was when they heard it. It is such fun giving people goose-pimples that we exaggerate the story – just a little, to make it a bit more colourful – and then the next person to pass the story on exaggerates a bit more, and so on. For example, having woken up to find that a famous person had died in the night, you might make enquiries to discover exactly when she died. The answer might come back, 'Oh, it must have been *approximately* 3 a.m.' Then you work out that you could well have been dreaming about her *somewhere around* 3 a.m. And before you know where you are, the 'approximately' and the 'somewhere around' get left out of the story as it does the rounds until it becomes: 'She died at *exactly* 3 a.m., and that is exactly the moment when my cousin's friend's wife's granddaughter was dreaming about her.'

Sometimes we can actually pin down the explanation of a weird coincidence. A great American scientist called Richard Feynman tragically lost his wife to cancer, and the clock in her room stopped at precisely the moment she died. Goose-pimples! But Dr Feynman was not a great scientist for nothing. He worked out the true explanation. The clock was faulty. If you picked it up and tilted it, it tended to stop. When Mrs Feynman died, the nurse needed to record the time for the official death certificate. The sickroom was rather dark, so she picked up the clock and tilted it towards the window in order to read it. And that was the moment at which the clock stopped. Not a miracle at all, just a faulty mechanism.

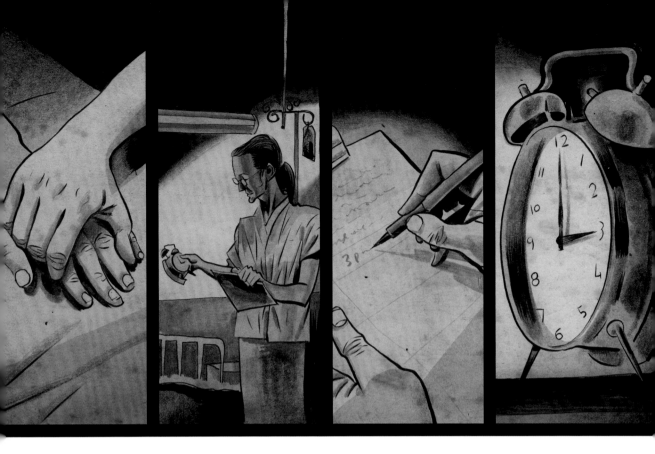

Even if there had been no such explanation, even if the clock's spring really had wound down to a stop at exactly the moment when Mrs Feynman died, we shouldn't be all that impressed. No doubt at any minute of every day or night, quite a lot of clocks in America stop. And quite a lot of people die every day. To repeat my earlier point, we don't bother to spread the 'news' that 'My clock stopped at exactly 4.50 p.m., and (would you believe it?) *nobody died.*'

One of the charlatans I mentioned in the chapter on magic used to pretend he could restart watches by the 'power of thought'. He would invite his large television audience to go and fetch any old broken-down watch in the house and clutch it in their hand while he tried to start it remotely with the power of thought. Almost immediately the phone in the studio would ring, and a breathless voice at the other end would announce, in awed tones, that their watch had started.

Part of the explanation may have been similar to that in the case of Mrs Feynman's clock. It's probably less true of modern digital watches, but in the days when watches had springs, simply picking up a stopped watch could sometimes restart it as the sudden movement activated the hairspring balance wheel. This can happen more easily if the watch is warmed up, and the heat from a person's hand can be enough to do that – not often, but it doesn't have to be often when you have 10,000 people, all over the country, picking up their stopped watches, perhaps shaking them, and then clutching them in warm hands. Only one of the 10,000 watches has to start in order for the owner to phone through the news in great excitement and impress the entire television audience. We never hear about the 9,999 watches that didn't restart.

A good way to think about miracles

There was a famous Scottish thinker in the eighteenth century called David Hume who made a clever point about miracles. He began by defining a miracle as a 'transgression' (or breaking) of a law of nature. Walking on water, or turning water into wine, or stopping or starting a clock by the power of thought alone, or turning a frog into a prince, would be good examples of breaking a law of nature. Miracles like that would be very disturbing indeed to science, for the reasons I gave in the chapter on magic. Disturbing *if* they ever happened, that is! So how should we respond to stories of miracles? This was the question Hume turned to; and his answer was the clever point I mentioned.

If you want to know Hume's actual words, here they are, but you have to remember that he wrote them more than two centuries ago, and English style has changed since then.

No testimony is sufficient to establish a miracle, unless the testimony be of such a kind, that its falsehood would be more miraculous than the fact which it endeavours to establish.

Let me try to put Hume's point into other words. If John tells you a miracle story, you should believe it only if it would be even more of a miracle for it to be a lie (or a mistake, or an illusion). For example, you might say, 'I would trust John with my life, he *never* tells a lie, it would be a *miracle* if John ever told a lie.' That's all well and good, but Hume would say something like this: 'However unlikely it might be that John could tell a lie, is it really *more* unlikely than the miracle that John claims to have seen?' Suppose John claimed to have watched a cow jump over the moon. No matter how trustworthy and honest John might normally be, the idea of his telling a lie (or having an honest hallucination) would be less of a miracle than a cow

literally jumping over the moon. So you should prefer the explanation that John was lying (or mistaken).

That was an extreme and imaginary example. Let's take something that really happened, to see how Hume's idea might work in practice. In 1917, two young English cousins called Frances Griffiths and Elsie Wright took photographs, which they said were of fairies. Above is one of their photographs, of Elsie posing with her 'fairies'.

You might think the photograph is an obvious fake, but at the time, when photography was still quite a new thing, even the great author Sir Arthur Conan Doyle, creator of the famously un-foolable Sherlock Holmes, was taken in by it, and so were quite a lot of other people. Years later, when Frances and Elsie were old women, they came clean and admitted that the 'fairies' were nothing more than cardboard cut-outs. But let's think like Hume, and work out why Conan Doyle and the others should have known better than to fall for the trick. Which of the following two

possibilities do you think would be the more miraculous, if it were true?

1 There really were fairies, tiny people with wings, flitting about among the flowers.

2 Elsie and Frances were making it up, and faking the photographs.

It's really no contest, is it? Children play make-believe all the time, and it is so easy to do. Even if it were hard to do; even if you felt that you knew Elsie and Frances very well, and they were always completely truthful girls, who would never dream of playing a trick; even if the girls had been given a truth drug, and had sailed through a lie-detector test with flying colours; even if this all added up to its being a *miracle* if they told a lie, what would Hume say? He would say that the 'miracle' of their lying would still be a *smaller* miracle than the fairies they claimed to show actually existing.

Elsie and Frances didn't do any serious harm with their prank, and it is even rather funny that they managed to fool the great Conan Doyle. But such tricks by young people are sometimes no laughing matter, to put it mildly. Back in the seventeenth century, in a village in New England called Salem, a group of young girls became hysterically obsessed with 'witches', and started imagining, or making up, all sorts of things which, unfortunately, the very superstitious adults of the community believed. Numerous older women, and some men too, were accused of being witches in league with the devil, and of casting spells on the girls, who said they had seen them flying through the air, or doing other strange things that witches were popularly believed to do. The consequences were extremely serious: the girls' testimony sent nearly twenty people to the gallows. One man was even ceremonially crushed under stones, which is an appalling thing to happen to an innocent person, purely because a group of children made up stories about him. I can't help wondering why the girls did it. Were they trying to impress each other? Could it have been a bit like the cruel 'cyber-bullying' that happens today in emails and on social networking sites? Or did they genuinely believe their own tall stories?

Let's come back to miracle stories in general, and how they get started. Perhaps the most famous instance of young girls saying weird things and being believed is the so-called miracle of Fatima. In 1917, at Fatima in Portugal, a ten-year-old shepherd girl called Lucia, accompanied by her two young cousins, Francisco and Jacinta, claimed to have seen a vision up on a hill. The children said the hill had been visited by a woman called the 'Virgin Mary', who, though long dead, had become a kind of goddess of the local religion. According to Lucia, the ghostly Mary spoke to her and told her and the other children that she would keep returning on the 13th of each month until October 13th, when she would perform a miracle to prove she was who she said she was. Rumours of the

expected miracle spread around Portugal, and on the appointed day a huge crowd of more than 70,000 is said to have gathered at the spot. The miracle, when it came, involved the sun. Accounts of exactly what the sun is supposed to have done vary. To some witnesses it seemed to 'dance', to others it whirled round and round like a Catherine wheel. The most dramatic claim was that

> . . . the sun seemed to tear itself from the heavens and come crashing down upon the horrified multitude . . . Just when it seemed that the ball of fire would fall upon and destroy them, the miracle ceased, and the sun resumed its normal place in the sky, shining forth as peacefully as ever.

Now, what do we think really happened? Was there really a miracle at Fatima? Did the ghostly Mary really appear? Conveniently, she was invisible to everybody except the three children, so we don't have to take that part of the story very seriously. But the miracle of the moving sun is supposed to have been seen by 70,000 people, so what are we to make of that? Did the sun really move (or did the Earth move relative to it, so that the sun appeared to move)? Let's think like Hume. Here are three possibilities to consider.

1 The sun really did move about the sky and come crashing down towards the horrified crowd, before resuming its former position. (Or the Earth changed its rotation pattern, in such a way that it looked as though the sun had moved.)

2 Neither the sun nor the Earth really moved, and 70,000 people simultaneously experienced a hallucination.

3 Nothing happened at all, and the whole incident was misreported, exaggerated or simply made up.

Which of these possibilities do you think is the most plausible? All three of them seem pretty unlikely. But surely Possibility 3 is the least far-fetched, the least deserving of the title of miracle. To accept Possibility 3 we only have to believe that somebody told a lie in reporting that 70,000 people saw the sun move, and the lie got repeated and spread around, just like any of the popular urban legends that whizz around the internet nowadays. Possibility 2 is less likely. It requires us to believe that 70,000 people simultaneously experienced a hallucination involving the sun. Rather far-fetched. But however unlikely – almost miraculous – Possibility 2 may seem, even that would be far less of a miracle than Possibility 1.

The sun is visible all over the daylight half of the world, not just in one Portuguese town. If it really had moved, millions of people all over the hemisphere – not just those in Fatima – would have been terrified out of their wits. Actually the case against Possibility 1 is even stronger than

that. If the sun really *had* moved at the speed reported – 'crashing down' towards the crowd – or if something had happened to change the Earth's spinning sufficiently to make it look as though the sun had moved at that colossal speed – it would have been the catastrophic end of all of us. Either the Earth would have been kicked out of its orbit and would now be a lifeless, cold rock hurtling through the dark void, or we'd have careered into the sun and been fried. Remember from Chapter 5 that the Earth is spinning at a rate of many hundreds of miles per hour (1,000 mph if measured at the equator), yet the apparent motion of the sun is still too slow for us to see it, because it is so far away. If sun and Earth suddenly moved relative to one another fast enough for a crowd to see the sun 'crashing down' towards them, the real movement would have to be thousands of times faster than usual and it literally would be the end of the world.

It was said that Lucia told her audience to stare at the sun. This is an extremely stupid thing

to do, by the way, because it could permanently damage your eyes. It also could induce a hallucination that the sun was wobbling in the sky. Even if only one person hallucinated, or lied about seeing the sun move, and told somebody else, who told somebody else, who told lots of other people, each of whom told lots of other people . . . that would be enough to start a popular rumour. Eventually one of those people who heard the rumour would be likely to write it down. But whether or not that's actually what happened is not what matters, for Hume. What matters is that, however implausible it might or might not be for 70,000 witnesses to be wrong, it is still far less implausible than for the sun to have moved in the way described.

Hume didn't come right out and say miracles are impossible. Instead, he asked us to think of a miracle as an improbable event – an event whose improbability we might estimate. The estimate doesn't have to be exact. It's enough that the improbability of a suggested miracle can be roughly placed on some sort of scale, and then compared with an alternative such as a hallucination, or a lie.

Let's go back to that game of cards I talked about in the first chapter. You remember we imagined that four players were each dealt a perfect hand: pure Clubs, pure Hearts, pure Spades, pure Diamonds. If this actually happened, what should we think about it? Again, we can write down three possibilities.

1 There has been a supernatural miracle, perpetrated by some wizard or witch or warlock or god with special powers, who violated the laws of science in such a way as to change all the little hearts and clubs and diamonds and spades on the cards, so that they were perfectly positioned for the deal.

2 It is a remarkable coincidence. The shuffling just happened to produce this particular perfect deal.

3 Somebody has performed a clever conjuring trick, perhaps substituting a previously doctored pack of cards which he had concealed up his sleeve, for the pack we all saw being shuffled out in the open.

Now, what do you think, bearing in mind Hume's advice? Each of the three possibilities may seem a bit hard to believe. But Possibility 3 is by far the easiest to believe. Possibility 2 could happen, but we have calculated how unlikely it is, and it is very very unlikely indeed: 53,644,737,765,488,792,839,237,440,000 to 1. We can't calculate the odds against Possibility 1 as precisely as that, but just think about it: some power or force, which has never been properly demonstrated and which nobody understands, manipulated red and black printing ink on dozens of cards simultaneously. You might be reluctant to use a strong word like 'impossible', but Hume isn't asking you to do that: all he's asking you to do is to compare it to the alternatives, which in this case

53,644,737,765,488,792,839,237,440,000 to 1

consist of a conjuring trick and a gigantic stroke of luck. Haven't we all seen conjuring tricks (often involving cards, by the way) which are at least as mind-boggling as this? Obviously the most likely explanation for the perfect deal is not pure luck, still less some miraculous interference with the laws of the universe, but a trick by a conjuror or a dishonest card-sharp.

Let's look at another famous miracle story, the one I mentioned earlier about the Jewish preacher called Jesus turning water into wine. Once again, we can list three main kinds of possible explanation.

1 It really happened. Water really did turn into wine.

2 It was a clever conjuring trick.

3 Nothing of the kind happened at all. It is just a story, a piece of fiction, that somebody made

up. Or there was a misunderstanding of something far less remarkable which really did happen.

I think there is not much doubt about the order of likelihood here. If Explanation 1 were true, it would violate some of the deepest scientific principles we know, for just the same kind of reason we met in the first chapter when talking about pumpkins and coaches, frogs and princes. Molecules of pure water would have to have been transformed into a complex mixture of molecules, including alcohol, tannins, sugars of various kinds and lots of others. The alternative explanations will have to be very unlikely indeed, if this one is to be preferred over them.

A conjuring trick is possible (much cleverer tricks than that are done regularly on stage and on television) – but less likely than Explanation 3. Why bother even to suggest a conjuring trick, given the lack of evidence that the incident occurred at all? Why even think about a conjuring trick, when Explanation 3 is so very likely, by comparison? Somebody made up the story. People invent stories all the time. That's what fiction is. Because it is so very plausible that the story is fiction, we don't need to trouble ourselves to think about conjuring tricks, still less about real miracles that violate the laws of science and overturn everything we know and understand about how the universe works.

As it happens, we know that lots of fiction has been made up about this particular preacher

called Jesus. For example, there's a pretty little song called the Cherry Tree Carol, which you may have sung or heard. It's about when Jesus was still inside his mother Mary's womb (that's the same Mary as in the Fatima story, by the way), and she was walking with her husband Joseph by a cherry tree. Mary wanted some cherries, but they were too high on the tree and she couldn't reach them. Joseph wasn't in the mood to climb trees, but . . .

> *Then up spoke baby Jesus*
> *From in Mary's womb:*
> *'Bend down, thou tallest branch,*
> *That my mother might have some.*
> *Bend down, thou tallest branch,*
> *That my mother might have some.'*

> *Then bent down the tallest branch,*
> *Till it touched Mary's hand.*
> *Cried she, 'Oh, look thou, Joseph,*
> *I have cherries by command.'*
> *Cried she, 'Oh, look thou, Joseph,*
> *I have cherries by command.'*

You won't find the cherry-tree story in any ancient holy book. Nobody, literally nobody who is at all knowledgeable or well educated, thinks it is anything but fiction. Plenty of people think the water-into-wine story is true, but everybody agrees that the cherry-tree story is fiction. The cherry-tree story was made up only about 500 years ago. The water-into-wine story is older. It appears in one of the four gospels of the Christian religion (the Gospel of John: none of the other three, as it happens), but there is no reason to believe it is anything but a made-up story – just one made up a few centuries earlier than the one about the cherry tree. All four of the gospels, by the way, were written long after the events that they purport to describe, and not one of them by an eye witness. It is safe to conclude that the water-into-wine story is pure fiction, just like the cherry-tree story.

We can say the same thing about all alleged miracles, all 'supernatural' explanations for anything. Suppose something happens that we don't understand, and we can't see how it could be fraud or trickery or lies: would it ever be right to conclude that it must be supernatural? No! As I explained in Chapter 1, that would put an end to all further discussion or investigation. It would be lazy, even dishonest, for it amounts to a claim that no natural explanation will ever be *possible*. If you claim that anything odd must be 'supernatural' you are not just saying you don't currently understand it; you are giving up and saying that it can never be understood.

Today's miracle, tomorrow's technology

There are things that not even the best scientists of today can explain. But that doesn't mean we should block off all investigation by resorting to phoney 'explanations' invoking magic or the supernatural, which don't actually explain at all. Just imagine how a medieval man – even the most educated man of his era – would have reacted if he had seen a jet plane, a laptop computer, a mobile telephone or a satnav device. He would probably have called them supernatural, miraculous. But these devices are now commonplace; and we

know how they work, for people have built them, following scientific principles. There never was a need to invoke magic or miracles or the supernatural, and we now see that the medieval man would have been wrong to do so.

We don't have to go back as far as medieval times to make the point. A gang of Victorian international criminals equipped with modern mobile phones could have coordinated their activities in ways that would have looked like telepathy to Sherlock Holmes. In Holmes's world, a suspect in a murder case who could prove that he was in New York the evening after the murder was committed in London would have a perfect alibi, because in the late nineteenth century it was impossible to be in New York and in London on the same day. Anyone who claimed otherwise would seem to be invoking the supernatural. Yet modern jet planes make it easy. The eminent science-fiction writer Arthur C. Clarke summed the point up as Clarke's Third Law: *Any sufficiently advanced technology is indistinguishable from magic.*

If a time machine were to carry us forward a century or so, we would see wonders that today we might think impossible – miracles. But it doesn't follow that everything we might think impossible today *will* happen in the future. Science-fiction writers can easily imagine a time machine – or an anti-gravity machine, or a rocket that can carry us faster than light. But the mere fact that we can imagine them is no reason to suppose that such machines will one day become reality. Some of the things we can imagine today may become real. Most will not.

The more you think about it, the more you realize that the very idea of a supernatural miracle is nonsense. If something happens that appears to be inexplicable by science, you can safely conclude one of two things. Either it didn't really happen (the observer was mistaken, or was lying, or was tricked); or we have exposed a shortcoming in present-day

science. If present-day science encounters an observation, or an experimental result, that it cannot explain, then we should not rest until we have improved our science so that it can provide an explanation. If it requires a radically new kind of science, a revolutionary science so strange that old scientists scarcely recognize it as science at all, that's fine too. It's happened before. But don't ever be lazy enough – defeatist enough – to say 'It must be supernatural' or 'It must be a miracle'. Say instead that it's a puzzle, it's strange, it's a challenge that we should rise to. Whether we rise to the challenge by questioning the truth of the observation, or by expanding our science in new and exciting directions, the proper and brave response to any such challenge is to tackle it head-on. And, until we have found a *proper* answer to the mystery, it's perfectly OK simply to say, 'This is something we don't yet understand, but we're working on it.' Indeed, it is the only honest thing to do.

Miracles, magic and myths – they can be fun, and we have had fun with them throughout this book. Everybody likes a good story, and I hope you enjoyed the myths with which

I began most of my chapters. But even more I hope that, in every chapter, you enjoyed the science that came after the myths. I hope you agree that the truth has a magic of its own. The truth is more magical – in the best and most exciting sense of the word – than any myth or made-up mystery or miracle. Science has its own magic: the magic of

reality.

Index

Acknowledgements

Richard Dawkins would like to thank:

Lalla Ward, Lawrence Krauss, Sally Gaminara, Gillian Somerscales, Philip Lord, Katrina Whone, Hilary Redmon; Ken Zetie, Tom Lowes, Owen Toller, Will Williams and Sam Roberts from St Paul's School, London; Alain Townsend, Bill Nye, Elisabeth Cornwell, Carolyn Porco, Christopher McKay, Jacqueline Simpson, Rosalind Temple, Andy Thomson, John Brockman, Kate Kettlewell, Mark Pagel, Michael Land, Todd Stiefel, Greg Langer, Robert Jacobs, Michael Yudkin, Oliver Pybus, Rand Russell, Edward Ashcroft, Greg Stikeleather, Paula Kirby, Anni Cole-Hamilton and the staff and pupils of Moray Firth School.

Dave McKean would like to thank:

Christian Krupa (computer modelling); Ruth Howard (Chemistry adviser), Andrew Hills (Physics adviser) and Cranbrook School; Clare, Yolanda and Liam McKean.

Picture credits

Galaxies, p. 167, © NASA/Getty

Spectroscope, p. 170, © Museum of the History of Science, Oxford

Spider, p.199, ©Thomas Shahan

Earthquake simulation, p. 206, ©The US Geological Survey and the Southern California Earthquake Center

Michael Jackson in car bonnet, p. 248, © KNS News

'Jesus in a frying pan', p. 249, © Caters News

'Jesus in toast', p. 249, © Chip Simons/Getty

Cottingley fairies, p. 255, © Glenn Hill/SSPL/Getty